D1809749

Climate Change, Religion, and Our Bodily Future

Studies in Body and Religion

Series Editors: Richard M. Carp, Saint Mary's College of California

Studies in Body and Religion publishes contemporary research and theory that addresses body as a fundamental category of analysis in the study of religion. Embodied humans conceive of, study, transmit, receive, and practice religion, with and through their bodies and bodily capacities. Volumes in this series will include diverse examples and perspectives on the roles and understandings of body in religion, as well as the influence and importance of religion for body. They will also move conversation on body and religion forward by problematizing "body," which, like "religion," is a contested concept. We do not know exactly what religion is, nor do we know exactly what body is, either; much less do we understand their mutual interpenetrations. This series aims to address this by bringing multiple understandings of body into an arena of conversation.

Recent Titles in Series:

Climate Change, Religion, and Our Bodily Future, by Todd LeVasseur
Creative Encounters, Appreciating Difference: Perspectives and Strategies, by Sam Gill
Religion and Technology into the Future: From Adam to Tomorrow's Eve, by Sam Gill
Sensing Sacred: Exploring the Human Senses in Practical Theology and Pastoral Care, edited by Jennifer Baldwin
Body of Christ Incarnate for You: Conceptualizing God's Desire for the Flesh by Adam Pryor
Sacred Scents in Early Islam and Christianity, by Mary Thurlkill
Dancing Bodies of Devotion: Fluid Gestures in Bharata Natyam, by Katherine C. Zubko
Early Daoist Dietary Practices: Examining Ways to Health and Longevity, by Shawn Arthur
Risky Marriage: HIV and Intimate Relationships in Tanzania, by Melissa Browning

Climate Change, Religion, and Our Bodily Future

Todd LeVasseur

LEXINGTON BOOKS
Lanham • Boulder • New York • London

Published by Lexington Books
An imprint of The Rowman & Littlefield Publishing Group, Inc.
4501 Forbes Boulevard, Suite 200, Lanham, Maryland 20706
www.rowman .com

86-90 Paul Street, London EC2A 4NE, United Kingdom

Copyright © 2021 by The Rowman & Littlefield Publishing Group, Inc.

All rights reserved. No part of this book may be reproduced in any form or by any electronic or mechanical means, including information storage and retrieval systems, without written permission from the publisher, except by a reviewer who may quote passages in a review.

British Library Cataloguing in Publication Information available

Library of Congress Cataloging-in-Publication Data

Names: LeVasseur, Todd, author.
Title: Climate change, religion, and our bodily future / Todd LeVasseur.
Description: Lanham : Lexington Books, [2021] | Includes bibliographical references and index. | Summary: "This book investigates how human-induced global warming will influence the bodily practice, performance, and production of religion in various geographic locations in the years and decades to come"—Provided by publisher.
Identifiers: LCCN 2021022611 (print) | LCCN 2021022612 (ebook) | ISBN 9781498534550 (cloth : alk. paper) | ISBN 9781498534567 (electronic)
Subjects: LCSH: Religion and geography—Forecasting. | Climatic changes—Effect of human beings on. | Global warming.
Classification: LCC BL65.G4 L48 2021 (print) | LCC BL65.G4 (ebook) | DDC 201/.77—dc23
LC record available at https://lccn.loc.gov/2021022611
LC ebook record available at https://lccn.loc.gov/2021022612

♾™ The paper used in this publication meets the minimum requirements of American National Standard for Information Sciences—Permanence of Paper for Printed Library Materials, ANSI/NISO Z39.48-1992.

This book is dedicated to two beautiful bodies:
Dr. Ajani Ademiwa "Ade" Ofunniyin
1952–2020
Colleague, mentor, inspiration, priest, artist, teacher,
shaman, creator, freedom fighter, prince, king, and now,
ancestor. You were and will forever be woke.
Shannon Horning
Wannabe chess master, in utmost gratitude for your years of friendship,
your beaming smile, and for all you have done for me, my family,
our community, and this sweet earth.

Contents

Acknowledgments

This book has been percolating over many years in the background of my life as other personal and professional obligations and opportunities have competed for my time and attention. Given the long gestation period of its writing I have many people to thank.

At Rowman and Littlefield, a variety of editors have helped the project. To Michael Gibson, Trevor Crowell, Becca Beurer, and Sarah Craig I offer thanks for your assistance and patience. I also offer thanks to the reviewer whose insights helped me clarify the flow of the book.

I was fortunate to be able to share some of the thoughts in the book at invited talks in England and thank for their hospitality and friendship Arran Stibbe at the University of Gloucestershire and Pasi Heikkurinen at the University of Leeds/University of Helsinki.

At the College of Charleston, I thank my inspiring and engaging colleagues and the many student interns at the Center for Sustainable Development. I also graciously thank the faculty fellows of the Sustainability Literacy Institute and their commitment to using education to make the world a more sustainable place: Steve Jaume, Dave Hansen, Caroline Foster, Barry Stiefel, Deb Bidwell, Kendall Deas, Jen Wright, and Leslie Hart. I also thank the many student, staff, and faculty members on the Quality Enhancement Plan's Implementation Committee for their support in my directing of the QEP project. A big thanks to Zeff Bjerken (and Luke Shirley) for getting me to Ladakh—hopefully we can do it again. A thanks as well to Gibbs Knotts, Dean of the School of Humanities and Social Sciences for his flexible leadership and support. Friendships with Anthony Greene, Renard Harris, Josh Bloodworth, Lisa Young, Leroy Lewis, Charissa Owens, Kris De Welde, Simon Lewis, Allison Welch, Merrie Koester, Kenyatta Grimmage, Blake Scott, Dan Dickison, and all in the Friday hoop crew have made my life better and I am thankful for their dignified presence on campus. Lastly, and to me, most importantly at CofC and thus in my life, a huge heartfelt thanks goes to my soil and fungi and grubnola and dolmas-loving nihilist partner in crime who is also about 65 percent banana in his DNA, Seth Pritchard, and as well a thanks and a magical hug to all the students in our co-taught fall 2019 HONS course, who along with the students in my fall 2016 Spirit of Sustainability course, all give me hope for a better future.

Writing for me is an emotional process of creation, no less when on such a heavy topic as climate destabilization. My creative process is in large part fueled by background music that fits the mood of what I am putting to paper as I move through writer's block and synthesize research and sources. I thank the players of instruments who will be there who provided the soundtrack to my years of staring at a screen while working on this project: Bob Marley and the Wailers, Peter Tosh, Satsang, the Grateful Dead, Rage against the Machine, Rising Appalachia, Prince and the Revolution, Burning Spear, Trevor Hall, Michael Franti and Spearhead, Xavier Rudd and the United Nations, Alpha Blondy and the Solar System, Midnite, Akae Beka, and Ziggy Marley and the Melody Makers.

For the amazing cover that brings to flesh the trajectories of the book, I thank Nikki Scioscia. Her engaging and unique talent is matched by her desire to make the world a better place and I encourage the reader to visit her website and support her art: http://www.nikkiscioscia.com/.

For grounding friendship that included lots of laughs and adventures and support in times of struggle, I thank: Adam Cloyes, Jeremy Tunstill, Burton Callicott, Teak Smith, JR Yarnall, Ward Buckheister, Reva and Ary Fun, Spencer Higgs, Clayton Thomas-Muller, Garrett Boudinot, Chad Houfek, Crafton Dicus, Daniel Scruggs, Andrew Mendosa, Elliott Bazzano, Andrew Anastassiou, Lucas Johnston, Tanner Crunelle, Suzanna Ellison, Ian Schlieffer, Patsy Hallen, June McDaniel and Jim Deno, Luckey and Alina Thompson, Don Elhussein, Daniel Ostad, Vincent Schachner, Mike Price, Chris Adams, and Christa and Jason Hebal. For hosting regenerative visits I thank Nika, Farmer, Zev, and other collaborators up at Earthaven. Equal thanks to my family (both Halberdas/Hoffmans, and also LeVasseurs/Tischlers), and especially Mike for the many COVID-19 Facetime calls and to my dad for letting us dwell on Folly.

Paul Pulé has provided a grounding in care for which I am thankful, while Alastair McIntosh has regaled me with stories of fairies while reminding me to be open to the deeper mythopoeisis of life (while also pouring me my first and only shots of authentic Hebridean Scottish whiskey). And for years of continued letters of support and constant professional encouragement I offer humble and deeply appreciative thanks to both Anna Peterson and Bron Taylor.

My last and deepest of thanks go to two sets of bodies. The first are the bodies of Jeanette, Aviva, and Leo—what a blessed life to share it with you three, who gift endless meaning and depth to my existence. The other is to a singular body, that of my editor Richard Carp, who was open to my original vision for his edited series and that resulted in this book. I am forever thankful to his gracious patience, professional demeanor, insightful comments, and deft nudges that helped this volume become more than I had hoped it could ever be. His guidance in thinking and feeling through the many limitations of Settler Colonialism and the Academy in a time of climate chaos were instrumental for my own voice that emerges in the pages that follow, and it is to him, his friendship, and his editorial gifts that I offer my endless gratitude.

Preface: Where to, Next, in Our Bodies, with "Climate Warming?"

In his 2006 book *The Creation*, the famed biologist and secular humanist E. O. Wilson, in recognizing and reflecting on the importance of religion in people's lives, was prompted to write a fictitious pro-environmental letter to an imagined member of the Christian clergy. His thesis was that concern for planetary health should unite us all, whether one is an atheist (like him), or one is someone who takes the creation story of Genesis seriously (his imagined audience). As a nonspecialist, he would not know the difference between traditional nature-as-sacred religiosities, and religious environmentalist discourses (Tomalin 2013), but his thesis based on US (and moreso, global) demographics about religion are apropos. Most *Homo sapiens* are religious, and their identities—social, ethical, emotional, political, affective, and environmental—are impacted in significant ways by their religious identity, practice, and belief. What humans do with religion to shape their bodies, at individual and corporate levels, and what they do with religion to respond to a rapidly changing planetary environment driven by human-induced global warming, are the interlocked foci of this study.

I labor in this book to do two things. These dual goals are mutually reinforcing, although also at times through the text they are treated as unique to themselves:

1) Help contribute to, and in some small capacity, possibly add something new, or at least reimagined, to theoretical discussions in religious studies about religious bodies (i.e., a resurgent religious studies, flooded with the world[1]). This is the focus of the part I of this book through chapter 3, with this effort creating the entry into my next goal:

2) Model what theorizing about religious bodies during a time of climate crisis may look like, to be done in part II of the book in chapters 4

through 7, and explain why such theorizing is a necessary condition for the academic study of religion (which I do throughout the book). This goal can be broken into two interconnected framing mechanisms:

a) One frame that is bio-ecologically materialist—we are (em)bodied animals in ecologies of place, so that as these places change, how humans as animals live in material environments will change, too[2];

b) Another frame that is thus anthropo-bio-ecological—because we are embedded bio-ecological beings with culture, where central to most culture is religion, as ecosystems change due to global warming, so too will the practice of religion change and specifically, the embodied and thus performed practice of religion. ·

Taken together my focus in this book is on theorizing how changes brought by human-induced climate change/global warming/climate warming[3] (that includes potential runaway climate change) in the coming decades will alter the landscape of religion. Goal 1 provides the theoretical context for this exploration and is articulated in part I of the book; and goal 2 the application of this theorizing to selected case studies in part II of the book. In so doing I recognize that with goal 1 there exists a vast body of literature in the anthropology of religion that engages with bodies. Moreso, some of this engagement is at the interface of culture and environment. However, my intended audience in this book, the rest of the field of religious studies and theological studies, is not typically engaged in biocultural theorizing. This point leads to me to goals 2, 2a, and 2b.

Many humanities scholars still have not taken seriously the severe impacts that rapidly changing biogeochemical processes will have on cultural production. This is a key theme of this book and in its aid I will unpack some of the natural science-based concepts related to ecology, ecosystem flows, and biogeochemical cycles that provide both the literal and theoretical grounding for this theme. In short, to know bodies and religion, we need to know the bio-ecological systems[4] where such bodies reside. As these systems move towards stochastic bio-ecological regime shifts triggered by trillions of tons of greenhouse gases blanketing our atmospheric commons, our bodies are being and will continue to be brought along for the bumpy ride. Parts of this book will provide a cartography of such planetary and bodily materialisms as they are shaped by carbon dioxide (CO_2), methane, and other gaseous forces. Given the above goals and their implications, I argue that to do good religious studies scholarship that is theoretically capable and analytically informed will, from here on out, require knowing precisely such global warming materialisms.

The editor for this series on religion and the body, Richard Carp, in explaining its function writes that, "Embodied humans conceive of, study, transmit, receive, and practice religion, with and through their bodies and bodily capacities" (2012, front page). This book contributes to the series by asking what it means to be embodied animals who conceive of, study, transmit, receive, and practice religion on a planet undergoing rapid biogeochemical and bio-ecological shifts. It also moves beyond, by asking what value such understandings of material bodies may provide for a ruptural and transgressive praxis of higher education, especially for religious studies.

One of the key arguments articulated throughout part I of the book is that the future theorizing and study of religious bodies will have to take into account the larger bio-ecological context within which our religious-in-practice bodies perform religion. I maintain that such capable and informed scholarship in an era of climate change is refuge making.[5] It will liberate thought from European dualisms while at the same time being cognizant of the ongoing violence against certain material bodies brought by extractive fossil fuel economies grafted onto the Settler Colonialisms of the last 500 years. The academy is an offshoot of such extractive economics and structurally benefits from such trajectories and ongoing colonialisms and this book is in part a challenge to some of the assumptions underpinning large swathes of this academy: that faith in reductionism is warranted; that most applied research should be overdetermined by embedded technophilia; that there are things called "disciplines"; that scholarship is neutral; and that humans are somehow hermetically sealed from the maladaptive bio-ecological repercussions of industrial lifeways. But more, this "decentering of European logics is not just a theoretical exercise of decolonization but a realignment of sense through affective infrastructures, an affective mattering in the discourse of materiality and its worlds" (Yusoff 2018, 98). So while we bioculturally and dramaturgically perform this thing we call "academic work," we must recognize such work is just one of many possible things we can perform as we dwell in ecologies of place.

Furthermore, what this human-constructed academic performance will look like in the coming decades will shift as the earth's climate also shifts—hopefully the performance of scholarship will shift toward understanding materiality by in part decolonizing such academic performance from European logics of domination. Moreso, I hope the academy as a whole is able to shift to performing teaching and research that provides hope in the dark (Solnit 2005) while recognizing achieving sustainability is incredibly hard work (Schendler 2009). And as alarming and complex as the consensus science of global warming already is as articulated by the Intergovernmental Panel on Climate Change (McIntosh 2020), we must respond to this science by attempting to engage in pleasure activism (brown [*sic*] 2019) central to which is our performed efforts of teaching to issues that will become ever more overdetermined by global warming.

The two goals of this work combined with a motivation for such decentered refuge-making reflect my own professional concern that, as scholars of religion, we need to deepen our conversations about bodies and recognize the limits of some major theoretical approaches to this conversation that have occurred to date. To be blunt, but in so doing recognizing the loss of needed nuance and glossing over the many examples (such as the multiple monographs on the anthropology of religious bodies) that can be found, I nonetheless maintain that as an imagined and performed whole the field of religious studies has not taken our animal bodies and their biological embeddedness seriously enough. This book hopes to add to the recent re-emergence of the body in religious studies by thinking through anticipated impacts of climate change on our religious bodies.

These two goals of (1) analytically theorizing religion in an age of climate change in part I and (2) using case studies to model the value of interpreting the embodied practice of religion on a climate changed planet in part II also deeply reflect my own personal concerns. These are concerns with another type of refuge: where and how will humans live twenty to thirty to forty years out, and beyond? What lifeforms will be with us at that time, and in 2100 and beyond? Which ones will be lost, and for what reasons? Will the ills of racism, sexism, xenophobia, homophobia, transphobia, and many other -isms and phobias be amplified during the coming bio-ecological pinch points and stressors? Key here is an all too real concern about religious fundamentalism and its often violent policing of certain bodies, and how climate change stressors will very likely add to this dynamic, too.

These questions are layered onto the existential reality that climate change triggered by human actions is scary. The more I sit with unfolding scenarios and each updated study about real-time and predicted impacts of climate change, the more scared I become. This tone is present throughout the book, as in a way I am writing back to myself from 2060, asking my younger self (and academia) why I (and we) did not do more to stop the trends we see emerging from my (and our) vantage point of 2020? For example, a team of scientists have once again tried to sound the alarm in a 2021 publication, writing in the abstract for their article published in *Frontiers in Conservation Science*, "we review the evidence that future environmental conditions will be far more dangerous than currently believed. The scale of the threats to the biosphere and all its lifeforms—including humanity—is in fact so great that it is difficult to grasp for even well-informed experts . . . .We especially draw attention to the lack of appreciation of the enormous challenges to creating a sustainable future" (Bradshaw, et al, 1).

This tone of anguish, concern, and frustration is also informed by my current ecology of place from where I dwell and perform my academic work: where I type this is literally 500 yards from the Atlantic Ocean—my whole community is vulnerable. My current place of employment already actively floods at certain high tides with whole streets inaccessible to vehicular traffic and it's predicted that by 2050 such sunny day flooding will occur for almost half of the year. I will be a climate refugee if the models of the future are correct, and given rates of ice loss from 1994 to 2017, where ice is melting 57 percent faster in 2017 than in the 1990s (Slater et al. 2020), this refugee status may not be too far off. Such rate tracks to a minimum 16 inch sea level rise from ice sheets alone by 2100 compared to pre-industrial levels, so if it is not in my lifetime (born in 1976), it is very likely that my children (at time of publication ages 11 and 5) and all the students I currently teach will be refugees from this and similar coastal regions the world over.

This book is in many ways therefore a plea for an urgent ethical examination of the role and purpose of how we perform religious studies (and higher education, broadly) in ecologies of place. Given ongoing and future climate changes, what research questions must we ask as scholars, and what courses should we teach as animals-with-agency who inhabit a rapidly destabilizing home planet that will force us to confront such eco-biological (and emotional) vulnerabilities? I return to answering such questions in the book's conclusion, but here share my agreement with the religion and ecology and religious ethics scholar Willis Jenkins who posits, "So far within religious studies . . . climate change typically appears as a special object of study rather than as labile pressure across an entire field of inquiry" (Jenkins 2017, 70).

This book is intended to help apply scholarly pressure to the field of religious studies (and theology) to help us better confront climate change vulnerabilities, including how we performatively theorize and engage with the domain of our object of study, "religion." Given this, it is a work of engaged scholarship. The applied case study portion of the book ends by looking at self-reported prescriptive solutions that may help uncover what can motivate our species in embodied religious biocultural contexts to undertake prosustainable behaviors to adapt to climate change. These insights then inform my own conclusion in the book about the need for us to perform engaged public scholarship that aids efforts of resiliency and adaptation to climate change, including in the academy.

The dual goals of this book matter because religious cosmologies are fluid, contested, polysemous, and complex (Tweed 2008; McCutcheon 1997; Chidester and Linenthal 1995). In this book I utilize the following operational understanding of the category "religion": religion (and thus the embodied and emplaced human-produced dramaturgical systems of religious production and performance) consists of a cosmologically/mythically grounded understanding of humans in community in relation to agent/s—power/s—being/s believed or experienced to have communicated (and often to continue to communicate) with humans-in-community guidelines for living that relate to and are centered upon interactions with those agent/s—power/s—being/s. Such agent/s—power/s—being/s may be visible or invisible, active in the past or currently active, and on-planet or off-planet (or both). The communicated guidelines shape and create systems of ethics, doctrines of belief, and bodily dramaturgical practices that inform an understanding of self and of collective community. Such a definition is fairly open ended and allows for a wide variety of forms of religion, including the posthuman practices I investigate in chapter 7 on resiliency and religious bodies. Such a definition also recognizes the always ongoing contact, boundary, and exchange (Tweed 1997) of religious tropes between and amongst human communities.

Anthropological and sociological studies suggest that humans use religious cosmologies to conceive of, and thus shape, human bodies, and this process has historically been tethered to local resources. This tethering influences both how these resources are managed, but also influences human conceptions of the self/the body in relation to these resources (Rappaport 1979; Turner 1967; Ingold 2000; Posey 1999). Within this milieu, the Western intellectual tradition, broadly speaking, has tended to abstract the *logos* and thus ego (think here Descartes' abstracting out and privileging of rational, ego-based thought) out of the body and the ecologies of places where our bodies live, building this dualism both on Christian doctrines and also Renaissance and Enlightenment technologies and epistemologies (Harvey 1990; Vasquéz 2011). Such religio-cultural scaffolding and performance of this abstraction has profound implications for both religious bodies and human-nature interactions.

For example, leading scholar on Native American issues, Vine Deloria Jr., writes that the historic Christian focus on linear sacred time and doctrine of sin have combined to contribute to environmental destruction and the colonization of non-Christians; while indigenous religions and their focus on sacred space and community have historically fostered radically different conceptions and enactments of religious bodies and human-nature interactions. His own argument, based on emerging consensus science about climate destabilization and species extinction in the early 1990s, was that "There probably is not sufficient time for the non-Indian population to understand the meaning of sacred lands and incorporate the idea into their lives and practices" (1994, 281) in order to help trigger more sustainable human-earth relations. Deloria's key insight, echoed by many others, and with which this book will enter into dialogue, is that a changing environment will result in changing practices and performances (and thus, most likely, conceptions) of the human body—at individual, religious, social, political, and environmental levels. What a home planet without summer ice at the north pole that is losing half of extant species, and that is growing to over 9 billion humans, presages for religious bodily production is an urgent and pressing topic worth exploring. In fact, this is the most pressing topic given the threat multiplier that climate change presents, as it will compound and exacerbate all other liberatory concerns we may have as scholars, let alone as human beings. This book therefore strategically enters such exploration.

Deloria's claim regarding temporal scales is in part supported by Johan Rockström and his colleagues, who argue that "To meet the challenge of maintaining the Holocene state" in which extant religions have evolved a variety of planetary boundaries that "define the safe operating space for humanity with respect to the Earth system" must not be transgressed (2009, 472). Of these, two have been eclipsed: biodiversity loss and the nitrogen cycle, with some religious production already responding to the former. The British-based international anti-poverty group, Oxfam, in the lead up to the Rio+20 meeting held in June, 2012, equally recognized the importance of resilient ecosystem functioning, without which humans will live severely impoverished lives. They recognize that human management of five key synergistic social and biological boundaries will largely determine the human future on earth (2012). These include food, energy, income, carbon, and nitrogen, with this management related to inequalities based on gender, race, class, and place (Mallory 2013; Shiva 2009; Finney 2014). The entrenched inequalities in these areas impact human-nature interactions and will equally influence religious bodily production in ecologies of place, especially as we progress further into the Anthropocene.

A word here on this term, "Anthropocene." It was first used by the Dutch Nobel Laureate and atmospheric chemist Paul Crutzen. The purpose and intent behind his nomenclature is a recognition that human actions have become of such scale that they are shaping a new planetary regime, one that will be measurable as such millions of years in the geological future. Between overloads of plastic, the release of radioactive isotopes, extinctions, and most ominously, the emission of trillions of tons of greenhouse gases, humans have become a geological force now able to be measured as a planetary epoch. So long Holocene, hello Anthropocene! As can be expected, multiple criticisms of this term have emerged. Two are worth highlighting right now with the first from the professor of inhuman [*sic*] geography Kathryn Yusoff, who writes that

> The Anthropocene began with the annihilation of the Colonial Other and an epochal redescription of geography as Global-World-Space. That is, the fungibility [i.e., interchangeable units of a commodity—in this case, minerals, then flora and fauna, then coal, then oil, but underlying it all, black and brown and indigenous bodies] of Blackness and geologic resources (as land, minerals, and ores) is coeval, predicated on the ability of the colonizer to both describe and operationalize world-space as a global entity. In this spike, the colonial Other

is displaced, along with existing ecological relations and connections of the colonized to earth. As Global-World-Space is established by the colonizers, the Human and its Others [i.e., White Europeans as Human, with the other side of the binary as Others: resources to be extracted to benefit the Human. These resources are black, brown, and indigenous bodies, and also earth metals, ores, and flora and fauna] are bifurcated in the production of racial difference to create two worlds of colonizer and colonized . . . Black and brown death is the precondition of every Anthropocene origin story. (2018, 32, 66)

Attention to issues of violent racism and violent dispossession is required to think through religious bodies in climate change. Such a thread of attention will at times be present in the book, but when it is absent should not be taken as tacit acceptance of the violent racial underpinnings of the Anthropocene. Here, too, academia has too often been the wrong kind of refuge for silencing the dynamics mentioned by Yusoff while also being sadly complicit in them, and our explorations of prosustainable futures must not replicate this history.

The second immediate point on the Anthropocene hints at one of the analytical lenses I will develop in the first part of the book: material feminism. Here I turn to professor of new media and communications, Joanna Zylinska, who points out that what "troubles me about the Anthropocene narrative [are] the gendered mode and tenor of [it as narrative], with its messianic-apocalyptic undertones and its masculinist-solutionist ambitions . . . the apocalyptic narrative of the Anthropocene also has an ontological dimension: it brings forth a temporarily wounded yet ultimately redeemed Man who can conquer time and space by rising above the geological mess he has created" (2018, 15). I too am leery of the gendered dynamics of Anthropocene as narrative and concept, residing as it does in larger trajectories of agency-denying scientific discourses of women and the natural world, both (Merchant 1980). I agree further when Zylinska writes that "the idea that the separation of Man from Nature signifies teleological maturation underpins the tragic vision of the world" (2018, 34). This separation is one of the moves challenged by the prosustainable embodied religious practices I will explore in chapter 7. This separation is also thoroughly embedded in the academy, and thus influences the goals of this book: for global warming will not only change the embodied practice of religion from here on out, it will change the equally as embodied practices and performances of teaching and scholarship we too often think are immune from biogeochemical shifts.

Overall, a climate and resource-stressed planet will reshape a variety of ontological commitments. The professor of English Tim Morton argues that climate destabilization is an example of a "hyperobject," and explains, "Our continued survival, and therefore the survival of the planet we're now dominating beyond all doubt, depends on our thinking past Nature." We need to move beyond Romantic-tinged views of a reified, pristine "Nature," and rather adopt what Morton calls "ecological thought." This is "the thinking of interconnectedness . . . . It's a practice and a process of becoming fully aware of how human beings are connected with other beings—animal, vegetable, or mineral. Ultimately, this includes thinking about democracy. What would a truly democratic encounter between truly equal beings look like, what would it be—can we even imagine it?" (all quotes 2009, 16). To answer this question, we cannot abstract ourselves out of matter. We cannot remove ourselves from our earthly homes and animal bodies. And we can't perpetuate such abstractions in the academy, or in religious studies. Answering these questions of creaturely democracy (what I call respectful relationality) is attendant on all laboring and dwelling in the academy, from economics (currently unable to take seriously externalities, or to stop extreme disparities of wealth) and political science (currently unable to help create a truly democratic, equitable, just society of equal representation and flourishing), to media studies (currently unable to limit media consolidation and propaganda), to our own field of religious studies (e.g., struggling with helping the public understand the political-social dangers of fundamentalisms). And all fields need to better integrate the insights from our colleagues undertaking liberatory labor in women's and gender studies, critical race studies, indigenous studies, and migration/immigration studies. If we cannot better ask and answer these questions and model compelling answers in the academy it will be that much harder for our species to approach the adaptation and resilience needed to equitably dwell in the emerging bio-ecologies of the Anthropocene.

NOTES

1. I draw from the work of Kathryn Yusoff here when using the phrase "flooded," with inspiration deriving from this specific passage: "Addressing origin stories is not just about making an alternative or alt-anthro-scene. Rather, it is to be attentive to what histories of the earth provide a break in analysis and narratives of material relations and languages of description that have colonized it, and to begin to make histories that launch a praxis for an insurgent geology into being—an insurgent geology that is . . . flooded with the world" (2018, 22).

2. The science of ecology makes "predictions about three types of objects: species, individuals, and traits or consequences of individuals. Parts of ecology (e.g., ecosystem theory) also make predictions about fluxes and pools of elements and energy . . . those fluxes and pools are controlled or affected by the activities, abundances, and distributions of organisms. Thus, they are aggregate consequences of species, individuals, or the traits of individuals" (Scheiner and Willig 2011, 10–11). As we are a species with religious traits, engaging in terraforming of the earth via industrial lifeways, our activities, abundances, and distributions will be impacted by such bio-ecological

fluxes. We are driving them, and as individuals and as a species will be impacted by
them. We must therefore study them by asking: how will such bio-ecological fluxes
impact the bodily practice of religion?

3. "Climate warming" is a phrase Australian environmental philosopher Glenn
Albrecht uses (2019), combining climate change and global warming into an apt
phrase that quickly gets beyond climate denialism.

4. At times I will also call these "ecologies of place" to recognize that these places
are shaped by our mammalian culture and dwelling practices which are part of ecol-
ogy/ecosystems. Ecologies of place allows for a more open-ended, emergent conception of places to occur; one where the culturally influenced actions and embodied
dwelling of human animals are part of the evolving and emergent bio-ecologies of
earth.

5. Here I build on the concept of refuge as shared by Stefano Harney and Fred
Moten in *The Undercommons: Fugitive Planning & Black Study* where they write,
"To enter this space is to inhabit the ruptural and enraptured disclosure of the commons that fugitive enlightenment enacts, the criminal, matricidal, queer, in the cistern,
on the stroll of the stolen life, the life stolen by enlightenment and stolen back, where
the commons give refuge, where the refuge gives commons" (2013, 28).

Introduction

Cascading changes in the interlinked biosocial regimes related to not only anthropogenic climate change, but also a variety of other environmental metrics (Morton's "hyperobjects"), are going to result in a very different human future where these changes will impact religious production, and thus human bodies, in a variety of sobering ways (LeVasseur 2015). Yet some of these bio-ecological changes will trigger religious changes in ecologies of place where some religious production by humans-in-places will respond to crossing tipping points in creative, dynamic ways. Some humans will increasingly bio-culturally perform and enact religious behaviors to help (or at least attempt to help) adapt to climate change and other environmental stresses (Taylor 2010; Bauman 2014; Grim and Tucker 2014; Johnston 2013; McDuff 2010; Globus Veldman et al. 2014), with evidence of this presented later in this book. Such adaptive responses provide a ray of hope, as some of this production will contribute to an overall "ecology of human flourishing" (Swearer and McGarry 2011) built on more respectfully relational encounters with nonhuman others, mediated via religious production and performance.

The overall structure of this book tracks the above and engages with these issues as follows:

Chapters 1, 2, and 3 make up part I and combine to undertake a literature review from various disciplines that provide an academic framework for analyzing the case studies and theoretical claims introduced in the subsequent chapters of part II. These three chapters in part I are where I grapple with the first guiding research question of this book: how can scholars better theorize religious bodies in the context of human-induced rapid climate change? Chapter 1 introduces an exploration of the bio-ecological future, and what this means for religious bodies that are always emplaced in ecologies of place. It briefly contextualizes the study of religion, the body, and nature in the academy. It undertakes a critique of religion-body theorizing, paying attention to how the more-than-human world has been silenced and largely made absent thus leaving us as scholars underprepared to better theorize ongoing and emerging biocultural responses to shifts in global climates. Biocultural materialist insights are brought to the fore to help think through bodily impacts of current and future global warming on bio-ecologies of place. I briefly explore leading causes of unsustainability that are driving climate change and how these maladaptive developments may be viewed through both a philosophy of place and a dwelling perspective. This chapter grounds the "object" (via object-oriented ontology) of climate change and how global warming will inform our "day-to-day, spiritual, emotional, sexual, social, cultural and political lives [as these] are conducted in relation to objects and thoroughly mediated by them" (Candlin and Guins 2009, 1–2). This especially includes objects derived from naturally occurring biogeo-chemical processes that are now in the beginning of a chaotic shift to new regimes. Humans will undergo solastalgia[1] as the planet we were born on, and the neighborhoods and bio-ecosystems and ecologies of place we grew up knowing and interacting with, radically shift as a result of a warming of 2–4 degrees Celsius this century. Changes from global warming mean that everything will be in flux and that the cultural and built infrastructure of the past will be severely impacted by new climate regimes (what Bill McKibben calls "eaarth") and the performance and production of religion by individual and corporate bodies will not escape this reality. I introduce in this chapter one of the guiding analytical lenses of the book, as well: religious bodily dramaturgy. This concept is the thread that stitches my arguments together throughout the entire book.

Chapter 2 provides a brief overview of the deep evolutionary history of our individual bodies as any understanding of bodies must recognize that "nature" presents limits on what our bodies can do, perform, and accomplish. This scientific backgrounding will be used to make the case that this deep evolutionary past will inform and impact human responses to the environmental future, and to future religious production, as it has always done. Here I discuss micro-, macro-, and thus meso-level evolution, and how biocultural religious dramaturgical flexibility will occur, within parameters, at the meso-level. The chapter recognizes that my approach to investigating the human body is built upon a certain techno-scientific discourse, one that is helpful in understanding our bodies; however, other religious knowledge systems about bodies exist, and Hindu conceptions of the body will be shared in order to recognize heterogeneity in conceiving of bodies via religions, consistent with the larger aims of this edited series. A key insight in chapter 2 will be that an "anthropology of the senses" (Seremetakis 1994, 4) that takes "discussions of the sensuous body" (Stoller 1997, xv) seriously, must emplace our sense-bound biophysical creaturely bodies within a planet undergoing climatic shifts. These shifts will impact the embodied performance of biocultural religious dramaturgies as they occur in bio-ecological places.

Chapter 3 summarizes thoughtways from the first two chapters. It revisits the concept of nature and then looks at nature and bodies through key analytical lenses: queer theory, materialism, feminist new materialism, posthumanism, and ecological animisms. It also looks at concepts of agency, and which bodies and material forces have agency,[2] and how these thus impact the performance of religious bodily dramaturgies in bio-ecological places.

With this literature review in place that grounds the analyses of religious bodies offered in this book I move to part II of the book. It is in part II of the book that I engage with my second guiding research question (and subquestions) and undertake case studies that model the type of theorizing about religious bodies during an era of rapid climate change I am arguing the field needs to undertake. To accomplish this in chapter 4 I discuss an overdeterminate material agency: oil. Here I explore the politics and built environments of "petrocultures," looking at how the quest for coal and then especially oil has contributed to the climate change of our time. To understand religion and the body, we must understand oil cultures.

Chapter 4 is a "pivot" chapter for the volume after which I move even further into the key case studies that both model the type of scholarship for which I advocate in this book and that help better situate my arguments with clear examples of religious bodily dramaturgies that are responding to climate change. Chapter 5 brings a performance studies lens to the dramaturgies of river worship in India, where climate change is rapidly melting the glaciers that feed the Ganges (Ma Ganga) and Yamuna (Ma Yamuna) Rivers, impacting the embodied worship of both; and to the site of the *hajj* in Saudi Arabia, where climate change is rapidly heating up the area. Chapters 6 and 7 build on these explorations by analyzing preliminary evidence for prosustainable religious dramaturgies that are aimed at more resilient and adaptive biocultural lifeways in bio-ecological places. Chapter 6 journeys to Ladakh, India, and emerging Buddhist dramaturgies of bio-ecological places, and chapter 7 encounters post-materialist Global North nature-as-sacred biocultures where at the meso-level practitioners are trying to create adaptive dwelling practices.

The volume concludes with a call for a new type of education praxis for the public scholar. It is a call that takes material agencies and posthuman ontologies as its assumed starting point to help the performance of scholarship and teaching in the academy better adapt to a hotter, warmer, shifting future of life that is already here. This call is required because many scholars, myself included, are wedded to a tragic vision that too many of us perpetuate in our scholarship and our teaching. This is the tragic vision that sees our species, and thus, ourselves, as separate. All too often this inherited vision guides the performance of our professional practices in our respective ecologies of place, and we pass this vision on to our students. Such tragic vision delays the needed moves toward adaptation and resiliency that the academy must aid and for which it must advocate, especially if it (and we in religious studies and theology) is to remain relevant. It is my hope that this book can provide a resource to aid religious studies, theological studies, and the environmental humanities more broadly, in advocating for such adaptive resiliency in how the academy is conceived and performatively structured in the bio-ecological, material places where we dwell.

The approaches I utilize in this book are intended to help move scholarship in the academy, broadly, and religious studies, specifically, to an agenda of postanthropocentrism. They are also intended to aid scholars in "drawing cartographies of the power operational in and immanent to the production of discourses and practices circulating in our sociopolitical order and integral to our subject formation" (Braidotti 2017, 83). In this book this task includes understanding and analyzing oil-based techno-political-ecological energy flows in an era of climatic chaos and how this will impact bodily enactment of religion.

For example, what happens when we take seriously the following insight from Australian environmental philosopher Glen Albrecht, who writes, "As the threat to humanity of a powerful climatic *global* force of our own making has never before existed, much of what has been written in the past about nature and life is irrelevant to the future we now face" (2019, x). Given this insight, I argue that we are fugitives, in need of a commons: a queered commons that is gender, sex, race, species, and every other intersectional identity just and equitable; a commons that "unsettles dominant imaginaries" (May 2015), including imaginaries that abstract human animals out from the matter that is everything, and that thus justifies exploitation of earth bodies and the bodies of certain humans based on logics of domination. Academia, and within it, religious studies and theology, need to rupture origin stories of Isolated Humans acting in this thing called "the economy," where the earth is (and certain bodies are) lesser, simply inert materials to be manipulated and commodified. We need a praxis attuned to the harsh reality of the Anthropocene, and for what comes next. A praxis attuned to, as per insights from object-oriented ontology, what is already here. We need a praxis of engaged scholarship and teaching (King 2015) attuned to the global force of our own making. So while I graciously applaud the many works that address intersectional power dynamics in religious contexts—one of the needed and key developments in the field over the last 30–40 years—these still fall short. The threat multiplier of climate warming is only going to make these power differentials worse, and all too often such work fails to address this reality. Rather, from the vantage point of 30 to 40 years from now, our praxis must recognize that most of what has been written and taught to date in the academy (especially an academy based on the material logics of petroculture), and for sure in religious studies and theology, will largely be irrelevant to the future we now face.

Given our respective commitments (often subconscious) to anthropocentrism, abstraction, and bio-ecological distancing, as well as emotional and psychological attachment (often conscious) to professional training and identity (and thus attachment to the larger capitalist system within which academia exists), some readers may immediately react defensively to my claims. However, my goal is not to criticize specific academics, but the broader system in which we all are performatively engaged. Most academics have exerted tremendous efforts to gain their achievements, following a system of rewards not of their creating, along predetermined pathways that are largely inflexible (such brittleness and lack of adaptive learning is part of the problem with the academy, and also religious studies). Yet pressures of "publish or perish" and the overall structure of academic work and our assumptions about it reflect the colonization of academic institutions and of how we perform academic work. To date most work and the institutions within which it occurs perpetuate logics that are ill-equipped to address the rapid collapse of bio-ecological systems the world over; the same bio-ecological systems within which our academic work always performatively occurs. I ask readers to please see this book as an offering along the way of liberating all of us from an outdated, flawed academy that is dependent on larger trajectories of violence that impact material bodies in bio-ecological places, and that all too often, mostly without honest reflection, aids in perpetuating the violence against these same bodies.

Our to-date focus as a field, and in academia, broadly, is symptomatic of larger problems consonant with themes of this book and the instrumental focus of our efforts is facing a literal dead end. My own instrumental efforts are guilty of this, for example, every time I fly to a conference to give a 20-minute talk, or to India with students—this is materially helping to tip us further into such a dead end. Or every time I teach a course that does not invite students into having honest discussions about their bio-ecological futures. Or every time I do not challenge the hegemony of short-term or anthropocentric thinking at my institution, or I am complicit with supporting general education requirements that speak nothing to the unfolding Anthropocene.

Let me clarify this with an example from the field of religious studies at large. Throughout the year I receive various pamphlets from a wide variety of academic presses, many of them historically esteemed presses where if you publish with them you are in the fast track for tenure and promotion. Roughly speaking, a vast majority of these publications from learned and highly capable scholars have nothing to do with the material reality of a planet passing through biogeochemical tipping points. And roughly speaking most of them have nothing to do with helping us better understand this moment of bio-ecological fracturing, or with helping us better strategize and generate resilient and adaptive lifeways attuned to these fracturing places. They do not mention sustainability, or wicked problems, or resilience, or any of the issues grappled with in this book.

Instead I see in these pamphlets, and in the key trade journals of religious studies, in colleagues' archival research, or textual research, or ethnographies of self-defined key categories of religion, or most presentations at conferences, or at the book exhibit at the annual American Academy of Religion (AAR) meeting, almost nothing to do with climate-change induced collapse. I see almost nothing to do with our creaturely embeddedness in evolutionary material bio-ecological flows and places. And in almost the entirety of all of these works, based on perusing the offerings from multiple publishing houses and conference proceedings going back years now, I see no effort to connect this work to the literal collapse of the bio-ecologies within which we are bioculturally embedded. I would even hazard a guess that most of us in religious studies who do grapple with these issues are at a loss to explain to our colleagues what we do and why it matters. And even if our colleagues "get" that environmental metrics are bad, for whatever reasons, they most likely will not let this inform their future respective teaching or research.[3] Which makes them, and thus all of us, and what we perform at a professional level in ecologies of place basically irrelevant to the larger bio-ecological reality of life in the years and decades to come. Which is the only reality that is, and thus that actually matters.

What knowledge formations can inform a praxis of dwelling in and on the reality of this, our climate changing home, besides those both utilized and explored in this book? For sure we must be leery of pinning too much hope on knowledge formations emerging from mainstream religions, as "religious environmentalism has little traction within the world's predominant religions [such that] the best available evidence indicates that [mainstream] religion is not coming to the environmental rescue any time soon" (Taylor 2017, 931). Those religions I will highlight in chapter 7 point to possible emergent knowledge formations, although for surfing, FuckForForest, and EcoSexuals such systems are still being formed from within petrocultures and their attendant imaginaries, violence, and biomaterial flows and biocultural Global North regimes of power. However, in terms of meso-level biocultural adaptive becoming, the historical record suggests that, by definition of longevity, those biocultures that have lasted the longest have ceremonial-based traditional ecological knowledge bases. Religious groups in the Global North dramaturgically moving in this direction should not only be studied, but taken seriously as steps along a path toward sustainable dwelling in bio-ecological places.

Scholar of indigenous resource management Fikret Berkes defines traditional ecological knowledge as *"a cumulative body of knowledge, practice, and belief, evolving by adaptive processes, and handed down through generations by cultural transmission, about the relationship of living beings (including humans) with one another and with their environment"* (1999, 8, italics in original). Note here this is a dramaturgy—an embodied and performed complex of knowledge, practice, and belief, attuned over time to bio-ecologies of place populated by living beings and agental material forces. Can we help shift the academy to forming and performing such knowledge systems in ecologies of place, and do so in a way that does not replicate colonial and neocolonial settler systems of power over, romanticization, and appropriation? Yet this also assumes that traditional ecological knowledge is only for indigenous peoples. This misses the point. Rather, all bio-ecological places where humans dwell are places of knowledge, practice, and belief. It so happens that the current knowledge-practice-belief systems of most human biocultural places privilege certain types of one species of living beings, key ways of thinking/belief, and economically rewards short-term resource extraction of places served by those maladaptive key ways of thinking/belief. Academia is largely designed to aid and perpetuate such maladaptive knowledge-practice-belief systems; while the actual physical and material places where academics work, college and university campuses, are as much constructed bio-ecological places where knowledge, practice, and belief are produced and performed as any other place where humans have and currently dwell.

Given this current design of academia, how can we shift epistemological fidelity to new models, to aid the development of new bio-ecological dramaturgies of place? This is a question we must attend to, to help navigate the ongoing reorganization of life around trillions of tons of greenhouse gases. These gases are here and their material agency will overdetermine biocultural lifeways and bio-ecologies of place for the rest of our tenure on earth. Academia must respond and adapt to this, and within academia, so must religious studies. Or both will cease to have survival value in future knowledge-practice-belief systems and ecologies of place.

The native scholar of indigenous education, Gregory Cajete, points to a possible path of place-based education that can aid such a shift. In concluding this Introduction I share this as its implications are that the academy is most likely set up to give exactly and entirely the wrong type of education, at least for the emerging Anthropocene. Cajete writes:

> American Indian education historically occurred in a holistic social context that developed a sense of the importance of each individual as a contributing member of the social group. Essentially, tribal education worked as a cultural

and life-sustaining process. It was a process of education that unfolded through reciprocal relationships between one's social group and the natural world. This relationship involved all dimensions of one's being while providing both personal development and technical skills through participation in the life of the community. It was essentially an integrated expression of environmental education. (2005, 69–70)

I know I was not trained to do the above. Nor do I come from a family or cultural history that can readily and easily replicate the above. And this holistic life-sustaining vision of education centered upon community, reciprocity, relationship building, technical skill building, and participation is surely not the one I am inspired or empowered to create or perform or participate in daily on a college campus. Most of us from petrocultures are in the same predicament. For example, the Citizen Potawatomi Nation botanist and founder of the Center for Native Peoples and the Environment, Robin Wall Kimmerer, shares her own experience of having the limits of her Western scientific botanical training being laid bare when she attended a conference on traditional ecological knowledge and heard a Navajo woman, one "without a day of university botany training in her life" (2013, 44) speak about plants. Kimmerer shares how the elder "spoke for hours [and] one by one, name by name, she told of the plants in her valley. Where each one lived, when it bloomed, who it liked to live near and all its relationships, who ate it, who lined their nests with its fibers, what kind of medicine it offered. She also shared the stories held by those plants, their origin myths, how they got their names, and what they have to tell us. She spoke of beauty" (2013, 44).

How do we bring beauty to the academy? How do we ask questions appropriate to the Anthropocene, so we can better understand how human animals are trying to protect such beauty, or are seeking to rediscover it? How can we help create institutions where knowledge of and ceremony with beauty are their reason for being?

We need a radical human ecology (Williams et al. 2012). A decolonial pathway forward (Mignolo and Walsh 2018). A radical reenchantment based upon the "intense desire to connect human life to the broader natural world [, not just in] the sciences but every domain of culture" (Gibson 2009, 148).

We need a ruptural (religious studies) scholarship of and about queer posthuman materialist dramaturgies in bio-ecological places, similar to what I model in this book. Such scholarship should support generating and nurturing a fugitive education for the commons of the same. This queer posthuman education of bio-ecological places must be attuned to material agencies and our sympoietic[4] co-creations with them. Our biocultural goals in this should be service, resilience, sheltering, empowerment, flourishing, grieving, justice, and ceremony.

It is toward these ends that this book is humbly offered.

NOTES

1. This concept is explored later in the book, but for a brief introduction here it is a term that relates to the felt, sensorial, psychological, and emotional experience of seeing places that were familiar become unfamiliar through the ongoing processes of climate-change shifts and unsustainable development. An example would be mountain top removal in Appalachia in the United States: locals witnessing the blowing up of mountain tops where they once hiked, fished, hunted, and camped. Despite the mountains being destroyed, the locals still dwell in the same ecologies of place where the mountains use to be which means that emotionally, affectively, and psychologically locals remember what used to be there while engaging with the vastly changed and impoverished present landscape, therefore creating a form of solastalgia. Add to this mountain top removal the climate changing weather patterns and the shifting seasonal migration patterns of insects and animals and changing patterns of flowers blooming: the lived experience of grappling with this change of place is solastalgia, grounded in memories of places that no longer exist as they once did.

2. My use of "agency" in this book implies the ability to choose how to respond to the material environment, and how to act within it. It therefore implies autonomy and self-direction, yet where for some agents these are influenced by culture, genes, and environmental pressures. It also implies forms of communication—a creaturely semiotics. For organisms agency is potentially strategic, and implied by action. Most (all?) biotic entities have agency. Yet I grapple with larger questions—if all is material, then can and do all material things have agency? For example, do ecosystems as a whole have agency? And another key question: do abiotic forces have not just effects, but also performative agency? I am struggling with this myself, as this is a foundational question about how we understand the world in which we are embedded: does all matter have agency? If so, how? If not, what matter does, and in what contexts? How would/do we know? Do greenhouse gases have effects (trapping sunlight energy as heat), or agency, or both? Right now I am swayed to accept the latter, but I am still very much wading through the various domains of literature that grapple with these questions and plan to pursue these questions in future publications. There is another foundational layer to this for scholars of religion: how does agency show up for "insiders" in religions, given my definition of religion, as compared to scholars of religion ("outsiders") who theorize about agental beings articulated in religious cosmologies?

3. These reasons may include disciplinary training/silos, and within these, advanced specialization; or lack of confidence in understanding climate science; or a recognition that if we have these honest discussions, then they may have to admit their jobs may be in jeopardy, too, or at least have to shift in focus, which may require new training.

4. Loosely speaking sympo- means "together," and poetic means "unfolding" or "creating" or "producing," so sympoeitic implies mutual co-creation.

Part I

THEORETICAL OVERVIEW

Chapter 1

Our Biocultural Future—
Whither the Environment?

Planetary Regimes and Bodily Immersion

A key goal for the first half of this chapter is exploring the larger bio-ecological, evolutionary context within which our bodies exist. For it is within this context, a context that when compared to geological baselines is one that is rapidly changing, that embodied religious production has always and will continue to occur. And it is this bio-ecological and biogeochemical evolutionary context that has been largely absent from the majority of our collective theorizing of religion and religious bodies.

For example, this seeming antipathy by some scholars of religion, who are willing to take bodies and embodiment seriously, but not take seriously the larger ecologies of place within which our animal bodies reside and have evolved, reared its head in 2014. Then acting president of the American Academy of Religion, Laurie Zoloth, had decided the theme of AAR's 2014 annual conference was to be on the environment and ecology with a special focus on climate change. Such a focus was decried as symptomatic of "the malaise poisoning the study of religion in the university," and this from a self-identified "life-long environmentalist" (Strenski 2014)! The charge by Ivan Strenski, a well-known figure in the field, was that this normative focus on climate change would pigeonhole scholarship and the questions religious studies scholars empirically and scientifically ask. Leaving aside that Strenski's own position is normative, it beggars the imagination to see how understanding the larger biogeochemical/bio-ecological context within which religious bodies operate does not fall within the purview of our field of study! How can this not be our purview when the scientific data is that such a context is becoming increasingly maladaptive for our survival fitness, and that this shift will impact our object of study: the human creation and performance of religion?

I am at a loss to see how we can undertake good scholarship, especially moving forward, if the larger biogeophysical context of scholarship within which our scholarship on religion is ignored, especially when that context is the actual place where people produce and perform religion. How we can undertake good teaching if we also collectively do not recognize that this biogeophysical context is shifting to new regimes, suggesting that traditional centers of learning[1]—universities—will have to adapt in both operations and content delivered to remain worthy of their respective mission statements? Rather, bodies are products of evolution in ecological places—this is the matter that makes them (and therefore us)—and this book is written precisely from this evolutionarily normative understanding. It is also written in light of religion and nature scholar Lucas Johnston's response to Strenski's critique of the AAR theme, when Johnston aptly pointed out that "religious studies has long embraced problem-oriented scholarship, from work on pluralism, to postcolonial studies, to queer theory, to critiques of racism and ethnocentrism. To defend the work of the academy as the disinterested and objective pursuit of knowledge about the world is to recapitulate the worst crimes of the academy" (Johnston 2016, 5).

While historically many scholars understand the body as a site of religious world building, many have stopped their focus on the body here. For reasons cultural and academic (and born of the culture of Western academia, broadly) most religion scholars do not extend their theoretical lens of bodies past the permeable membranes of our embodied biophysical collection of evolved cells to factor in the larger natural world. Unfortunately this myopia has blanketed our field for too long and left us unable to robustly and critically theorize religion on a planet passing tipping points (LeVasseur 2015b). The implication is that religion, if and when it does occur in (policed) bodies, ends there. We are still *cogitos*, but this time policed by regimes of power; or we are bodies that are gendered, worked, birthed, sick and healed, and eventually dead; or we are products of colonial and imperial and pole/metropole globalized trajectories and subaltern counter-trajectories. But one thing we are not in too much religious studies scholarship[2]:biologically evolved animals embedded and participating in natural habitats and ecologies of place.

My own scholarly habitat as an environmental humanist who focuses specifically on religion and nature interactions is emblematic of the return of the environment to certain humanities scholarship, however still in the minority of overall scholarship in the humanities, that has gained steam from the 1970s onward. We see this reemergence of the environment (and thus, for some, of place) in both the humanities and social sciences with the creation of environmental philosophy, environmental sociology, environmental anthropology, environmental economics, environmental law, and interdisciplinary environmental/sustainability programs of study the world over. Note here that the distancing term of "environment" is still used, implying a remove of the human, as if the environment is something out there. Better would be the label "ecological," which implies no center, no periphery, and no separation.[3] Regardless, the earliest of this move into what we now call the environmental humanities began in the 1970s, but most humanities and social science subdisciplines with an "environmental" as part of its nomenclature formed in the 1980s and 1990s as a response to the "Age of Ecology": the passing of environmental legislation in the 1970s; the first Earth Day of 1970; the publication of the 1987 "Brundtland Report" that operationally defined sustainable development; and the 1992 United Nations Rio Earth Summit. This era marked a new understanding of and concern for the Earth's biogeochemical systems, and both the perceived and scientifically tracked deleterious impacts our species was having on them.

Such understanding and concern have also informed the academic study of religion. This is seen, for example, in the creation of a religion and ecology group at AAR, as well as a religion and animals group, and more recently, a religion and food group. It is also seen in the work of Mary Evelyn Tucker and John Grim, who oversaw the series of conferences held at Harvard University throughout the 1990s and early 2000s on religion and ecology and who now oversee a Master's program in religion and ecology that is affiliated with the Yale Divinity School and Yale School of Forestry. It is also seen in the religion and nature PhD program at the University of Florida, from where I earned my PhD, and the ongoing publication of both *Worldviews: Environment, Culture, Religion* and *Journal for the Study of Religion, Nature and Culture*, the two leading academic journals devoted to the critical study of religion and nature/ecology.

Other key developments have helped create the context for the "reemergence" of the environment in humanities, and specifically religious studies, theorizing. One is the creation of the trope "lived religion" in the 1990s that helped bring the focus of religious practice back into the lived experiences of everyday peoples, including in their built and natural environments (Hall 1997; Orsi 1999; Gould 2005). The 1990s to early 2000s also saw the beginning of a sustained examination of how the environment informed religious views, practices, and beliefs, in first a largely North American but slowly international context (Lane 2001; Deloria 1994; Albanese 1991; Taylor 1995, 2005).

Another relevant development began in the 1970s and especially 1980s, helped by foundational work of Michel Foucault, Judith Butler, and Donna Haraway. The attention to the policing of (sexualized) bodies by regimes of power (discursive, legal, penal) (Foucault and Rabinow 1984) helped shape the academic understanding of bodies, and understanding of academic bodies, so they were no longer seen as blank slates or predetermined essences, but already produced sites of power dynamics. This is similarly echoed in Butler's work on critical gender theory that looks past essentialism to the social construction of gender and the politics tethered to construction of the female subject-as-object (1990). Meanwhile, Haraway (1996) helped explain the need for posthumanities scholarship by focusing on the already mingled human/technology (cyborg) symbioses at play in the twentieth century, and how understanding (or lack) of such symbioses was blinkered due to the patriarchal nature of science and science studies. All three of these scholars brought a needed focus back to embodiment and the political impacts of culturally dominant gender discourses. Their work also rightly points to the exclusionary practices of policed and performed subject creation within the context of patriarchal regimes of empire-based power.

While not alone in their respective projects, their voices and influence nonetheless loom large, past and present. A parallel and often overlapping movement also emerged in the 1970s and gained standing in the 1980s and 1990s: the development of the feminist care tradition of ethics, and feminist scholarship, broadly. In Australia key ecofeminist voices emerged that focused on binary dualisms tethered to a logic of domination. Here it was argued that the association of women's bodies with the natural world in patriarchal societies leads to the exploitation of both (Plumwood 1993; Warren 1990), to which ecowomanism adds that this exploitation is even worse for bodies of color (Harris 2017). Meanwhile, the independent scholar Carol Adams brought to animal philosophy a focus on silent female voices, agencies, and bodies and the dual exploitation of human female bodies, and animal other bodies. Her concern toward the latter was specifically in the domain of animal imagery and its impact on animal exploitation in dietary choices and industrial food regimes.

Together these scholars added to "the feminist recognition of interrelated oppressions" (Adams and Donovan 2007, 11) and helped bring to scholarship a focus on emotion, and always preexisting intersubjective relations based on embodiment, feelings, and care. Here the feeling, sensuous, sensual body is a legitimate site of meaning-making and world navigation, as compared to the disembodied, male, rational *cogito* of the Enlightenment project. The feminist scholar of history and science Carolyn Merchant adroitly argued that the latter has devalued the natural world, thus legitimating its exploitation (Merchant 1980). Similar developments of feminist scholarship emerged in the academic study of religion, as well, where "Feminist work on religion has participated in the multiform strategies taken up by feminist studies as a whole: historical recovery and reconstruction, imaginative reconstitution of traditions and practices, and ideological critique" (Castelli 2001, 4).

One of the most recent strands in this reemergence of the body in theorizing bodies and the natural world is the queering of ecology, where both the erotic as well as the natural world are theorized from beyond hetero-patriarchal binaries (Gaard 1997; Mortimer-Sandilands and Erickson 2010). Queering ecology means to not view the world through any essentialisms, including evolutionary ones, but to rather envision "liquid life [as] life-forms constitute a *mesh*, a nontotalizable, open-ended concatenation of interrelations that blur and confound boundaries at practically any level: between species, between the living and the nonliving, between organism and environment" (Morton 2010, 275–276). This highlights a key strand of inquiry and thought that runs through this book: how will religious bodies be impacted by these various interrelations and the collapsing boundaries wrought by climate change? As religion and nature/queer theorist scholar Whitney Bauman points out, the twenty-first century wicked problems[4] related to climate change that feature prominently in this book imply entanglement, multiple causality, and multiple understandings of time (2017). They therefore imply that where and when exploring human-animal body-religion-nature interactions such exploration must occur from multiple sites and sightings (2016). Entanglement means thinking about religion and bodies from multiple points of view, from multiple places of inquiry, from various domains of scale and time. It also requires moving away from linear cause-and-effect assumptions about human agency to realizing that we are embedded, biocultural beings constituted by and constitutive of ecologies of place.

Thinking this way can be aided by Indigenous Knowledges (IK) that can help to form a Western posthumanist ontology that explores a variety of human-nonhuman assemblages (Watson and Huntington 2008, 259) and encounters that occur in immanent spaces (Watson and Huntington 2008, 275). This means that in the move beyond the Cartesian project of abstracting a male-centered rational ego out of an intersubjective, messy, embodied real-ity (and other humanist and Enlightenment turns, including the privileging of objective, positivistic science based on controlled and replicable experi-ments), we need to be clear that some knowledge systems never went down this path. That in the rush to embrace re-enchantment, to dig in the ontological compostist[5] dirt of the earth, we do not want to perpetuate colonial categories of appropriation and power over. We need to recognize that "It is easier for Euro-Western people to tangle with a symbolic polar bear on a Greenpeace website or in a tweet than it is to acknowledge arctic indigenous peoples and their knowledge systems and legal-political realities" (Todd 2016, 6).

Rather, the intellectual lenses I bring to this book, including at times eco-logical animist insights, are offered in solidarity of critiquing "the European academy's *continued, collective* reticence to address its own racist and colo-nial roots, and debt to Indigenous thinkers in a meaningful and structural way" (Todd 2016, 10, italics in original).

Yet I must be ethically leery if I cross over into ontological spaces with which I am culturally unfamiliar. For example, scholar of Indigenous Peoples, Technoscience, and Environment Kim TallBear (Dakota) points out that "indigenous peoples have never forgotten that nonhumans are agential beings engaged in social relations that profoundly shape human lives. In addition, for many indigenous peoples, their nonhuman others may not be understood in even critical Western frameworks as *living*" (2015, 234). New animisms, object-oriented ontologies, posthumanism, queer inhumanities (Muñoz 2015), Afrofuturism, feminist new materialisms, ecological animisms, compostist terraworldings—all speak to ways of bodily natures that exist beyond the solipsism of Enlightenment, humanistic, patriarchal, CisHetero White Settler Colonialist thinking that permeates the historical past and too much of the present of the academy. The latter is ill-equipped to help attune human-animal biocultural bodies and lifeways with the bio-ecological places within which they reside. New ways (which for some are old ways) of thinking and being are in urgent need as we continue our inexorable climb to 500 ppm CO2 in our atmospheric commons and beyond.

Another point on the above insights from ecofeminism, ecowomanism, and IW: the bio-ecological system shifts of climate change will disproportionately impact and target Indigenous bodies and knowledge systems, as well as those of women, LGBQT+, black, brown, and peasant bodies and knowledge systems. The act of writing this book about the unfolding impact of climate changes on religious bodies is a form of privileged extraction that shelters myself, and most readers, from this reality. Add to this the impact of climate changes on the bodies of nonhuman others, and the future generations of human and nonhuman beings, and very quickly the potential for cheap appropriation of cli-fi (climate fiction) tragedies looms large. I hope to navigate this by bringing a respectful voice, attuned to the lived reality of suffering of those not at fault for climate changes, to the trajectory of this book.

I chart here a trajectory that recognizes our collective planetary future is going to be very different than the planet that our species has inhabited over the last 10,000 years of the now-ended Holocene.[6] This now-ended era is also the one in which all religions created by humans that are found on the planet today developed in (and for many of those, also began), as well. Because religion is embodied in human practices, and bodies inhabit material spaces that are both religious and not religious, then understanding bio-ecological shifts will help us better understand possible and likely shifts in religious bodies in the years and decades to come.

Here I want to be clear that I am a biological determinist at the macro and micro-levels: our bodies operate within evolutionary parameters at the cellular and individual levels (to be explored next chapter); as well as at the level of ecosystems and planet, which I explore here. So while there is flexibility in how we interact with our ecological environments, and in how we behave as individuals and collections of humans living together, the options available to us in our behaviors are constrained by existing evolutionary fiat.[7] This is not to say that we do not have social flexibility in how we behave as evolutionarily evolved primates. In fact, our inherent social flexibility at the meso-level, to be explored in chapter 3, leads to a variety of religious bodily practices, some of which are now responding (and will continue to respond) to larger changing environmental trajectories.

Another key point of clarity: there is no telos in evolution, and thus no anthropic principle animating the universe or the evolved forms of life on our shared planetary home. Rather, life forms evolve through various types of symbioses, environmental pressures and possibilities and adaptive mutations, and complex, interconnected interactions that continually occur. However, within this ongoing cacophony of the living state of all organisms, where there is always interdependent eating/consuming/birthing/intermingling, evolutionary changes can be rapid (Goodwin 1994). There is also a variety of systemic properties emerging at various scales, from the molecular and cellular to the planetary (Harding 2006). Some of these are also, while dynamic and interdependent, relatively stable in their relationship structure across scales of time and space.

These emergent properties at various scales are of import for this book, for certain climatic conditions, soil health, availability of fresh water, and other quotidian needs we have as primates are all thermodynamically and biogeochemically complex emergent physical attributes of our dynamic Earth.[8] Moreso: we are derivative biological organisms of these emergent planetary properties. And as these emergent biogeochemical contexts continue to shift, largely due at this point in geological time to human activity, we will be along for the evolutionary ride, wherever that may lead (Gunderson and Holling 2001).

For the immediate survival fitness of our species—and I write this as someone intrinsically concerned with the survival of the millions of other species we've evolved with and that are currently being driven extinct at rates that dwarf baseline extinction rates (De Vos et al. 2014)—we must be aware of key activities that are shifting our planetary home: our use of fossil fuels and our agriculture. Although this book focuses almost entirely on the former, these two activities are key drivers of change for the baseline Holocene biogeochemical patterns in which we have bio-ecologically evolved and in which our bodily religious future will occur.

Agriculture, the purposive domestication and utilization of species of flora and fauna to produce calories for human consumption, when scaled outwards around the planet is the older driver. Settled agriculture is typically harmful for soil health and leads to deforestation, making agriculture a driver of climate change by contributing to the release of carbon dioxide into the atmosphere. It also contributes methane, via rice farming (and the farming of other crops) and the use of domesticated animals (especially bovines) (Mason 2006[9]). There is also the burning of forests for pastureland which diminishes biodiversity and adds to global warming. These practices may be relatively benign over geological time when localized, but when enacted by hundreds of millions of human animals the world over can have a profound geological impact. This is especially true of industrial agriculture based upon Green Revolution technologies and intensive animal agriculture,[10] to the point where agriculture as practiced by most humans today is one of the biggest causes of anthropogenic climate destabilization[11] (FAO 2006; Manning 2005; Jackson 2011). Here modern industrial agriculture is a key hinge upon which petro-carbon extraction and use depends; in industrialized Green Revolution agriculture both key drivers of climate destabilization are present, as industrial agriculture is utterly dependent upon oil and natural gas derived products, which are actually bodies of old plants and sea life, as explored below.

Let's pause on the end of the above sentence: climate destabilization. The breakdown of the geological climatic patterns we evolved into—not just a change in climate, but destabilizing existing climate regimes so that entirely new climate regimes are now unfolding, triggered by human action. By tapping into ancient fossil reserves of decomposed plants and shells, and then manufacturing these ancient fossils into combustible forms of energy, the human-animal has released trillions of tons of once-stored carbon into our atmosphere as carbon dioxide. This project literally gained steam during the Industrial Revolution and is now global in scale today. The impact of carbon dioxide on our atmosphere, and one also mirrored by methane, is that as greenhouse gases their chemical compositions make them active in our atmosphere—they become material agents that actively determine the conditions of human bodily existence. This is because the material form of their chemical bonds have the effect of materially trapping ultraviolet light emitted by the burning of hydrogen in the sun. The internal space that carbon dioxide and methane are materially warming is the entirety of our planetary home, both terrestrial and aquatic.

The ongoing results, let alone the future predicted results, of this human caused global warming reads like a science fiction horror show: acidic oceans; species extinctions; shifting migrations of flora and fauna; changing weather patterns; more extreme weather events; thawing permafrost that will release more methane; increased desertification; new viruses entering into new vector territories; melting glaciers and ice caps; and sea-level rise.[12] From this point in time onward humans must adapt to this new planetary reality. This is because mitigation of most of these biogeochemical changes is not an option and currently no technologies exist to decarbonize the planet, even if such technology could be made to scale. And even implementing this technological-based adaptation requires a political and economic will that is currently in short supply: "the long-term scenario [of climate change] looks very serious, but the rate of change is glacial in terms of the typical politician's duration of office or most people's willingness to sacrifice for future generations" (McIntosh 2008, 40).

The author and climate activist Bill McKibben describes this new planetary home we have created as "Eaarth," where the interconnected feedback loops described above lead him to quip, "We need to dampen our intuitive sense that the future will resemble the past, and our standard-issue optimism that the future will be ever easier" (2010, 86). We must understand discontinuities, tipping points, stochastic regime shifts, externalities, and better plan our economic, ethical, and material habits of living to drawdown our collective impacts and stabilize, and possibly reverse, global warming (Thiele 2011; Hawken 2017; Alley 2011; Jamieson 2014; Broome 2012). However, we must accept that reversing global warming may not be thermodynamically possible—once the ice caps are gone, there's no getting them back and no control planet to which we can revert. The energy release of old stored sunlight into our atmosphere means new planetary systems within which we must adapt, let alone all the other species along for the ride. The implications of this are severe, including for academia (LeVasseur 2014; Orr 2016). And we must always attend to power differentials, in both who is historically complicit in creating this mess and who is disproportionately continuing it (Marks 2015; Shiva 2016).

Why does the above matter to how we conceive of and theorize religious bodies? It does because we are derivative organisms, living through rapid shifts in baseline biogeochemical cycles that we have triggered. If we are to survive as a species we will have to adapt rapidly to this bio-ecological reality. This means the production and embodiment of religion via our physical and social bodies will adapt, too. The implications of this on the coming biocultural reorganization that we will witness are staggering, and our scholarship must attend to this. This is why we need to create and perform a new praxis of liberatory scholarship that must help us reflect on these changes and then adapt to them.

Here I will provide one quick example of environmental/biological material overdetermination that we will have to adapt to and that will thus impact future social possibilities: the soil budget available to past civilizations directly determined which members of society were able to accumulate wealth and thus dominate others, and which ones could not (Kohler et al. 2017). These power dynamics have often been scaffolded into religious production, typically by the creation of often-male priestly classes in agricultural societies. This is a single factor scenario that influenced social possibilities: soil budget. How much more complex will the impact on societies be when there are synergistic bio-ecological regime shifts entailing multiple bio-ecological material factors, the world over? And if soil budget was that deterministic of power in the past, then what of today's ongoing soil loss and nitrogen deposition that impacts soil health? And then what happens when we scaffold onto failing soil health the impact to soil of the inevitable future global warming that we will not be able to escape or halt (Brown and Caldeira 2017)? What then, with these shifts, for religious bodies, as we all must eat? It's hard to have functional religious bodies without functional food supplies. But with the amplifying trends of climate change, will there even be dependable food supplies? And if soil budget strongly correlated to social power in the past, who will have the power in this even more complex future scenario of soil health, food distribution, access over the remaining stores of accessible fossil fuels, and social power?

Yet it is a linguistic sleight of hand to write about soil budget as if it is something that can be isolated. All of the global warming data that emerged just within the time it has taken to write this book only shows steady progression toward more maladaptive changes and catastrophic losses for soil budget and a wide variety of mixed biogeochemical cycles, patterns, and emergent bio-ecological systems. For example, the summer of 2019 brought record heat waves to much of Europe. The fall and winter of 2019 brought record drought and fires to almost the entirety of Australia (their summer). The winter of 2019–2020 in the Northern Hemisphere brought record heat, and in January 2020 heat waves to Antarctica during its summer. Yes, heat waves, in Antarctica. In the autumn of 2020 the ocean temperature where I live was 5 °F warmer into October than normal while November of 2020 broke US temperature records by being 15 to 20 °F warmer than normal. 2020 also saw a near-record hurricane season with named storms entering into the Greek alphabet, while scientists were highly alarmed that as of October, 2020 Arctic ice had yet begun to form for winter of 2020–2021 due to warm ocean and air temperatures. What is alarming is that we know that future years will be even warmer, making these records from 2019 and 2020 seem downright cool in comparison to what temperatures will be in 2040, let alone 2080. All this heat trapped by the material agency of the inert but materially active molecules of greenhouse gases are driving the climate feedback loops that materially will shape our bodily futures. During the writing of this book my own home of Charleston, South Carolina materially suffered from massive rain events—1,000-year downpours, now every few years, including rain-induced flooding in both fall of 2019 and early spring of 2020. And now studies suggest that by 2030, if warming isn't stopped at 2 °C, the oceans will collapse, with terrestrial ecosystems collapsing shortly thereafter in the 2040s.[13]

If all of that were not enough, in 2020 there has been the onset of the COVID-19 (Corona) global pandemic where most likely this virus jumped species in China. This event shut down large parts of the global economy, with people in many geographical areas of the globe quarantined in their homes for weeks and then months in March, April, May, and into June of 2020. It took this unexpected and tragic virus to help drop annual greenhouse gas emissions about 8 percent for the globe, although part of this drop was due to less use of coal in North America due to a warmer 2019-2020 winter.[14] It should be a sobering indictment of the utter lack of leadership—especially political—that total planetary emissions would most likely have continued to grow in 2020 without the virus. This, despite global leaders recognizing at the Paris 2015 climate agreement that humans must live by carbon-neutral means by 2050 to avert a likely 2 °C rise in temperature (Figueres and Rivett-Carnac 2020).

The materiality of the COVID-19 virus belied the fragility of our animal bodies and the brittleness of a linear, service-based global economy. It also showed the role religious bodies play in bio- and geosocial political locations. In the United States, many conservative evangelicals held in-person services, despite state-level mandates to not have social gatherings. Catholic churches held masses in some countries. Pakistan continued to have *salat* prayer in public, with those in attendance ignoring the strongly suggested six-foot social distancing protocols that were put in place to "flatten the curve" of those carrying the virus. On one level this suggests that our religious bodies will do what they may, regardless of a virus; so how much moreso with climate changes? If people will continue with "religion as usual" during the emergence of a highly contagious and deadly global pandemic, will they continue as usual within threatening climate changes? Or at least continue to do so until those changes no longer allow religious practices to occur in places they had?

And what of the function of religion as appeasing fears and anxieties of those unsure of the virus and concerned for their safety? For example, "Jesus is my antibody" slogans were present in 2020 at various protests against governor-mandated lock-downs in some US states. Such a confluence of conservative and fundamentalist Christianity coupled with States' rights, distrust of science, and distrust of government is found throughout the climate denial movement, as well (Globus-Veldman 2019). Will the same groups of people as expertly researched by religion and nature scholar Globus-Veldman continue to respond in similar ways as COVID-19 protesters did, as weather patterns continue to change and laws are passed to mitigate greenhouse gas emissions? To me, these are all compelling questions that scholars will hopefully investigate and address, including those in religious studies and the environmental humanities more broadly.

The two case studies in chapter 5 situate and explore comparable dynamics, in that to date Hindus and Muslims in the Himalayas and Mecca are not for the time being changing their dramaturgical practices, including in response to global warming, in their respective climate-changing places. The chapters after chapter 5 investigate the opposite: human animals with religion who claim to be changing their values and their behaviors in order to from their perspective actualize more resilient, regenerative biocultural practices within religious contexts. Akin to those front-line workers with COVID-19 who worked for the health of their communities, what may the same look like within religious production and performance, where those motivated by some type of religious identity put their bodies into action for sustainability and community health in the face of climate change?

However, to answer such questions about bodies and religion we must understand the bio-ecological container/s within which human-animal bodies reside. We must also attend to the power dynamics inherent to the interactions of these bodies, and how these dynamics are also impacted by ecoscape flows (LeVasseur 2015a; Asgard-Jones 2013). We also must attend to the absencing of past environmental knowledges of place due to solastalgia (Albrecht 2005), where the loss of such embodied memories and places will impact future bodily production. It is in this context that we must then also analyze religious bodies, as done throughout the second half of this book.

NOTES

1. Or as editor Richard Carp said to me when providing feedback on the first draft of this book: maybe we need learning to occur in de-centered places, so education can provide localized adaptive capacities that can respond to the shifting climate warmed bio-ecological places within which we dwell. I am in support of such a vision, as it makes sense that education should be in service of such climate resiliency that by definition will and can only occur granularly, in local bio-ecological places.

2. There are subfields other than religion and nature/ecology and religion and animals that do engage with evolutionary theory; see here especially the cognitive science of religion.

3. I thank my editor Richard Carp for this insight. And to be sure, some of these fields do at times adopt the prefix "ecological," as for example ecological economics (which is a different subfield than environmental economics).

4. Wicked problems are those that involve multiple stakeholders, are caused by many interacting systems that are outside of human control (or comprehensive human management), and that are intergenerational in their effects. Prime examples are institutional racism, institutional sexism, biodiversity loss, and anthropogenic climate change.

5. This is a term coined by Donna Haraway and will be explored more fully in the section "Theorizing Bodies and the Environment, or Material Bodies in the Academy, Redux" in chapter 2.

6. Geologists organize the history of the planet into various epochs, with these largely defined by extant atmospheric conditions, geographic locations of continents,

and the type of terrestrial and oceanic flora and fauna present during these varied conditions. The Holocene begins about 12,000 years ago when the Paleolithic era dominated by Ice Age cycles ended and the atmosphere entered a geological window of warming and glacial retreat. It is during the warming of the Holocene that all modern agricultural crops have been domesticated and permanent settlement of humans at large scales occurred.

7. Given time and space constraints in this book I steer away from debates and discussions related to transhumanism and imagined transhumanist technofutures.

8. Think here the laws of thermodynamics, including that energy can neither be created or destroyed; that metabolic energy derived from the sun powers our bodies; and for biogeochemical cycles think of various interconnected planetary cycles such as nitrogen cycling, hydrology and water flow, air currents, the chemical makeup of our atmosphere, the soil food web, animal migrations, the changing of the seasons, and geologic weathering to name just a few.

9. Domesication is a term Jim Mason coined that moves animal liberation discourses beyond critiques of nonhuman animal domestication and instead focuses on the systemic, hidden, and visible violent subjugation of animals, the willful massacre and extinction of animal bodies, and aggressive subjugation of animal agencies. He links this domesication to violence against black, brown, and indigenous bodies, and also to female bodies and agency.

10. The Green Revolution refers to post–World War II developments in agricultural technologies (especially the creation of dwarf hybrid varietals of wheat, cotton, and rice) and practices (use of chemicals for fertilization and pest control, increasing use of irrigation for water). Green in this context is not synonymous for sustainable or environmentally friendly. Intensive animal agriculture refers to concentrated animal feeding operations (CAFOs) where thousands of animals are raised in factory-like conditions (also called factory farms). Green Revolution farming techniques and technologies, which include CAFOs, are also referred to as "industrial agriculture."

11. Anthropogenic means human caused or human induced. This points to human action as being a key driver of recent observed (geologically speaking) changes in global climate, as compared to geological fluctuations that naturally occur. It is human action, directly related to our food production and use of fossil fuels, that has moved the planet out of the cyclical geologic patterns of past regimes of climate change.

12. All climate change discussion is based upon the internationally peer-reviewed and democratically organized work of the Intergovernmental Panel on Climate Change. See http://ipcc.ch/publications_and_data/publications_and_data_reports.sht ml for most recent reports (accessed December 29, 2017). For easy to understand explanation of climate science that covers chemistry, physics, geology, and biology, as well as deconstruction of global warming misinformation/skeptics, see www.skepticalscience.com (accessed December 29, 2017).

13. https://www.nature.com/articles/s41586-020-2189-9.epdf?author_access _token=xeIRKeSD_AQ2aWEwjFA-2dRgN0jAjWel9jnR3ZoTv0Mu1Ah8wVLy6bv XVvBoMP6OTTYrbG8GM0auUesx4pdEt--E3_subRo9vjNSWStNtN0yKVsw2QC GsxASBb6hI6U8FIoFS5VEmFuWHlokeEpRcw%3D%3D Accessed 4/28/20.

14. https://www.npr.org/sections/coronavirus-live-updates/2020/04/30/84830
7092/greenhouse-gas-emissions-predicted-to-fall-nearly-8-largest-decrease-ever?utm
_medium=social&utm_campaign=npr&utm_term=nprnews&utm_source=facebook
.com&fbclid=IwAR01JnA7a2IC4pvFiG7UGhcSLYrNpQNQbDlCJQ4HVZgmdK
-CWs2Zp4oxLRg Accessed 5/1/2020.

Chapter 2

Evolutionary Antecedents and Meso-Level Creativity

This chapter situates my exploration of bodies, religion, and anthropogenic climate change within the evolutionary past and present of our evolved bio-physical bodies. With this brief overview of our evolved bodies, I move in chapter 3 to how culture impacts our bodies, but does so always within the container of evolution articulated here.[1] Given space constraints and my own training this chapter can only be a cursory overview of key data points that will help frame the arguments to come. I am also reminded of a quip one of my graduate professors made upon seeing that Richard Dawkins published a book on religion and religious belief. This professor pointed out that they'd never have the intellectual arrogance to try to write a book in a field about which they knew nothing (i.e., writing a book for Dawkins' field), so why was it acceptable for Dawkins, without any sustained and advanced training in religious studies or theories about religion, to write a book on religion? I keep this statement in mind as I wade in the first part of this chapter into literature that is well outside the domain of my own disciplinary training. Yet, that's part of the challenge of re-thinking and re-theorizing the human animal on a planet undergoing climate changes[2]: we will have to grapple with domains of knowledge and literature that are unfamiliar to us. We will also fail in this process, so lots of humility and vulnerability is required. However, this is literature which we must try to gain a working knowledge of in order to have needed interdisciplinary discussions over generating solutions to the many interlinked problems facing our species. Furthermore, such grappling must occur in ways that do not reify science, and scientists, as prophets (Walsh 2013) of some techno-topic future where the humanities are largely seen as being supportive to the natural sciences (Slingerland 2008). Or where "Men Repair the World for Me," a fantastic chapter from Joanna Zylinska's apropos *The End of Man: A Feminist Counterapocalypse* (2018), whether this patriarchal repairing occurs in academia, science and technology,[3] politics, business, or the nonprofit sector.

This grappling by humanists with key concepts in the natural sciences must not be subservient to the natural sciences. This is one possible reading of Edward Slingerland as shared by Lisa Sideris (2017, especially pages 68–71). Slingerland himself writes, "If I have perhaps erred on the side of pushing humanists more to engage with the institutional 'Other,' it is because the biggest barrier to truly interdisciplinary dialogue and collaboration is not the supposed arrogance of Nobel laureates or white-coated custodians of scientific truth. The primary roadblock is the rigid, outdated, ideologically based resistance to empirical enquiry and to even mild forms of realism that continue to dominate the humanities. It is time to move on" (2008, 311). Here I agree with Slingerland's project to investigate how "We appear to have evolved preferences for particular types of sights, tastes, sounds, and sensations" (2008, 311) such that it is false, and epistemologically dangerous, to view everything about humans as a social construct and to theorize that culture determines everything about humans and physical reality is not important.[4] Rather, our choices as biocultural beings will always be impacted by, and in many cases overdetermined by, evolutionary constraints. On a planet undergoing rapid anthropogenic climate change that will impact the rest of our species' time on this planet these bio-ecological constraints will most likely become even more pronounced. We must engage with this reality as both scholars of religion, and with students and colleagues throughout academia.

THE DEEP PAST AND EVOLUTIONARY UNFOLDING

I accept as valid the scientific insights contained in an evolutionary worldview that posits that over billions of years on earth cells evolved, and have evolved ever more complexity, and interact in ways that lead to newer and novel forms of emergent complexity. Moreso, this process is ongoing and will always continue until the heat death of our planet in a few billion years and over that time will contain extinction events that see a huge loss of diverse life forms. The past, present, and future unguided and unscripted flowering forth of evolutionary life forms carries with it extinction and die-off events, no telos, random mutations, and adaptive fitness for better survival chances, all occurring within sympoietic and symbiogenetic[5] (Haraway 2016) unfoldings at scales cellular to ecosystem to planetary. Therefore it is clear that there is no anthropic principle guiding the universe, but rather "The universe is fine-tuned for life . . . as life creates the universe, not the other way around" (Lanza 2010, 93). Within this fine-tuning is the insight that "environments are as much the product of organisms as organisms are of environments" (Lewontin and Levins 2007, 32). All organisms co-create their habitats and ecologies of place as evolution unfolds—human animals contribute to, and are thus shaped by, such co-creation. Bio-ecologies of place are therefore a result of our species' interactions with them, as much as who we are as biocultural animals is shaped by the collectively created environments within which we always dwell. There is no one-way linear causality in any of this, nor any separateness of humans from the rest of the process of evolution's meandering yet constant unfolding.

This is the most accurate way we have to date of understanding the observable universe (and now some are suggesting possibly multiverses!) of which we are co-creative members. So while we have cultural flexibility, and increasingly access to some technologies that allow for some humans (especially wealthier ones) to physically interact with evolutionary antecedents in new ways, the reality is that there are "ground rules" to existence. To geological date these rules have shaped, literally, all possibilities available to us as biocultural individuals and collectives. They also will influence responses to climate destabilization. Lastly, because of these rules of evolution, climate destabilization is going to severely limit future possibilities of living as biocultural beings on this planet.

Yet between the biological determinism of microlevels (cellular to organismic) and macro-levels (ecosystems, bio-ecological systems, biogeochemical cycles, laws of thermodynamics, unfolding structure of the galaxy and universe), there is a "Goldilocks" middle ground where certain levels of predetermined and purposeful biocultural flexibility and performance exists. This is the more deterministically fluid realm of the culture aspect of biocultural evolution. The feminist philosopher of science Kate Soper hints at this meso-level of chastened flexibility when she writes, "Human beings, like all other living creatures, are determined by biology because they are embodied, moral entities with specific genetic endowments and a particular sexual anatomy and physiology. But relative to other animals, and in part by virtue of their specific biological evolution, human beings are biologically underdetermined when it comes to their modes of experience and response to these conditions . . . It is therefore important that ecological argument does not conflate the biological, cultural, and symbolic dimensions" (1995, 316) of human animals. This is because we use cultural and symbolic dimensions of our animality to navigate living on a complex, bio-ecologically indifferent home planet. The embodied and performed religious dramaturgy (including its symbolic dimensions) about which I write in this book is part of a larger biocultural response mechanism humans have that allows for some choices in how we dwell within the larger deterministic unfolding of life, and death. And as I explore in chapters 6 and 7, it may require us to frontload and strategically utilize in religious dramaturgical ways these dimensions of being human animals in larger bio-ecological contexts and do so in ways that may help generate prosustainable behaviors if we are to adapt to climate change with some hope of survival, let alone flourishing.

We Are Primates

One benefit of taking evolution seriously is that the scientific insights of reality that an evolutionary worldview provides can liberate those still enthralled with certain cultural epistemologies (including Western ones) from speciesist pathologies that constrain human self-understanding and freedom of expression (Stibbe 2012). This is especially true if these insights occur within the Precautionary Principle[6] and as part of an ignorance-based worldview (Vitek and Jackson 2008). Taking seriously these insights can also help remove those of us in speciesist cultural epistemologies from our species-appointed position of superiority. This is because it has become increasingly clear to ethologists in their study of animal emotions, cognition, and social interactions (and which has already been known by many human cultures) that nonhuman animals also have language, tool use, and culture. For many mammals their interactions also clearly include concepts of play and appropriate moral behavior (Bekoff 2006; de Waal 2016).

We know that we are social primates that evolved from apes, who as a collective are descendants of ape species that arose in the Miocene era about 20 million years ago. The *Homo* genus of which we are members evolved from this larger species tree around 2 million years ago (Dunbar 2016, 5–8). One of the defining features of our evolutionary lineage is bipedalism (Dunbar 2016, 8). The ability to walk upright helped our ancestors move about ancient African ecosystems where these ecosystems greatly impacted our evolutionary development (National Research Council 2010). For *Homo sapiens* a key early evolutionary development of our species lineage that allows for us to interact bioculturally at the meso-level with the ecosystems within which we are embedded is the evolution of our unique mind triggered by evolutionary increases in cranial capacities; followed by the formation of culture based on shared language (Dunbar 2016, 20–23; Cela-Conde and Ayala 2007, chapter 10) and shared materials and material processes. This implies that not only can we not survive without brains; we can also not survive without culture.

Besides the evolutionary parameters of our biological and chemical needs—cellular health and growth, certain levels of calories and potable water, procreation, blood pH and capillary flow, metabolism, functional organs, nerves and neurotransmitters, physical and emotional needs in the womb and post-birth through puberty, habitat fitness—there are also evolved genetic parameters that impact us. These are related to DNA, RNA, and genetic reproduction and the impact genes (approximately 10^6 to 10^9 genes in the genome, and about 30,000 in a human body, for billions of bits of genetic DNA in a human cell) have on our psychological dispositions, physiology, morphology, sex and gender expression, physical health, ability to learn, and overall cognitive functioning. And not to mention the almost infinite complexity of our brains, which have approximately 10^9 neurons! To this we can add the impact of bio-ecological factors (resource stress, pollution) on gene expression, and the emerging insights of the field of epigenetics as a whole, so that our evolved genetic parameters are symbiogenetically interactive with the rest of life. Overall, the literature on these areas are vast, with ever-emerging studies suggesting various interplay and impacts of biological, genetic, and chemical structures from the cellular to organismic levels of who we are.

As evolved social primates "pathways of neural and chemical activity" impact our respective individual cognition, physical development, psychological health and well-being, and body type and size, so that "Our total behavior is therefore a unique combination of more or less stereotyped subunits that makes behaviors look familiar" (Lewontin and Levins 2007, 57, both quotes). Yet, key to biologists Richard Lewontin and Richard Levins, the behaviors unique to the human animal are always in interaction with all aspects of the environment[7]—humans are not machines, or computers, and part of their project as natural scientists is to deconstruct the metaphorical language used to convey evolution that frames us in such ways. In this they are in concert with the posthumanist and new materialist scholars I cite throughout this book and the type of language I struggle to employ in capturing this reality of interactive, embodied terraworlding and composist[8] life and death becoming.

As scholars we must always remember and internalize the implications of what this means. We are not separate from the rest of the natural world. There is no dualism. We cannot abstract ourselves out of life and into our disembodied rational brains. It is patently false—both epistemologically and ontologically—to posit some *geist*, spirit, rational Archimedean point, or any other reified fallacy of misplaced human uniqueness that separates us from biophysical and bio-ecological reality (and worse, in so doing, views us as superior to the rest of reality, and immune to the laws of said reality). To this we must recognize and accept that our emergent consciousness is even "part of the physical logic of nature" (Lanza 2010, 124).

Accepting the above evolved biophysical reality in the understanding of our place on the planet, and therefore in thinking about religion and bodies, implies that there are limits. Cultural stories and cultural views of consciousness and our bodies that abstract us out of biophysical reality should be constantly policed and challenged, including where these occur in the academy. This is especially true if they contribute to maladaptive ways of behaving that can and do lead to species die off, loss of biodiversity, and human and nonhuman suffering, as for example with most current varieties of conservative and fundamentalist evangelical Christianity in the United States (LeVasseur and Zaleha 2019).

The context of this entire book is predicated on the above, and on remembering and therefore realizing that our evolved bodies (as well as those of most currently extant evolved organisms and species) have only biophysically interacted with and known an atmosphere within certain parts per million concentrations of key greenhouse gases, especially carbon dioxide and methane. Our evolved bodies have only interacted ·with a planet that has ice caps. Our evolved bodies have only interacted with and known a planet with an Amazon rainforest, which provides a certain amount of oxygen to our shared atmospheric commons. Our evolved bodies have only interacted with and known oceans with a certain level of pH. Our evolved bodies have only very recently come to know a home with nuclear radiation, persistent organic pollutants, plastic gyres in the oceans, and eutrophication from agricultural fertilizers. They have known a planet with the largest geological record of biodiversity and a biomass of insects that dwarfs the imagination, yet tragically where all of this fecundity is rapidly being lost by the unsustainable material impacts of certain human behaviors as these co-create new bio-ecological regimes.

Thus, any and all religious production, and any and all embodied and performed religious dramaturgical behaviors, have only interacted with and known these planetary realities, too. Understanding this is what it means to develop an ecologically aware religious studies methodology. As the religious capacities of the human-animal have evolved (Bellah 2011, xiv), and continue to evolve, they will by definition always occur within the confines of planetary and species evolution briefly described in this section. This grounding—both literal and figurative—of our understanding of religion on our home planet has been absent for too long from much of our field. This absencing of the biophysical context of everything human has resulted in theories of religion that are at times myopically blinkered and are arguably unequipped to meet the task of understanding religious bodies in this very different future that is already the present. Such a state of ill-prepare is unacceptable on professional levels as it perpetuates a religious studies, and the academy more broadly, populated by scholars that are unprepared to dwell in the looming crisis of climate change. We also need to come to terms with the very real disproportionate impacts this global warming will have on certain types of material bodies, and thus on regimes of power. If our scholarship is not actively speaking to this, it is incomplete scholarship, and may even be of dubious value. Especially if it perpetuates the types of abstractions and absencing I flag in this book, which frustratingly has been the case already for far too long.

THEORIZING BODIES AND THE ENVIRONMENT, OR MATERIAL BODIES IN THE ACADEMY, REDUX

Because we are not primary producers but are derivative products of preexisting trophic pyramids and habitats that are now in biogeochemical flux, we must come back to our bodies and rediscover the ecologies of place where our bodies always dwell. I wrote earlier that the body had been de-privileged in (Western) academia, and that this was a part of larger cultural trajectories. This does not mean that scholars have not discussed and theorized the human body (see this series, for example, and the many monographs from the field of anthropology); however, as locative places of meaning-making and embodied agency, bodies have been more absent than present in the Western intellectual tradition.

The abstracting out of humans from place is part of a larger whole, where place itself has been absent from this same Western intellectual tradition. Such sloughing off of place from epistemological and ontological developments can be traced back to the onset of Greek philosophy, with such sloughing continuing through Enlightenment thinking (Casey 1997). At this point Descartes' famous dictum that locates epistemology in (male) rational thought and ontology off planet (via the soul, that only humans have) created bedrock assumptions upon which natural theology morphed into the academy. As the phenomenologist Edward Casey explains,

> One of the abiding ironies of the situation is that early modern thinkers, by insinuating a forced choice between place and space, and then between absolute and relative determinations of space itself, thrust apart what had been constructively and unproblematically combined in previous thinking. . . . Only within the transcendental idiom provided by Kant at the end of the eighteenth century will there be an express effort to recombine the divergent directions of absolutism and relativism in one coherent framework. But a terrific price has to be paid for this act of recombination: space is no longer situated in the physical world but in the subjectivity of the human mind that formally shapes this world. Moreover, and as a direct reflection of this transcendental turn, any residual sense that place is importantly distinct from space will have vanished, with the result that place is given no attentive consideration. (1997, 136)

Thus in the academy bodies can be studied by the natural sciences, or by psychology, or public health, but embedding human-animal bodies into bioregional places (McGinnis 1998; Thayer 2003) where biocultural systems are constructed and enacted is largely absent in the academic study of religion to this day. The topological morphology of local places, where knowledge already resides and exists, and within which we are always present and shape and are shaped by, is background to the human-only drama upon which our collective academic gaze has been set.

Thankfully chinks in this edifice began to appear in earnest in the 1900s. Some were built upon Darwinian insights, and some upon Ernst Haeckel's coining of the term *oiecologie/oekologie* (later: ecology) in 1866. Aldo Leopold generates a land ethic (1949) in this slipstream, and Rachel Carson (1962) invites us to consider the impact of industrial chemicals on the bodies of birds and insects. She brings this further, showing how through bioaccumulation the same chemicals impact our own bodily health. In very broad strokes this brings us to the intellectual history shared in this chapter and upon which I journey and add to as I move forward in this book.

Here immediately I want to share what it is to think through religious bodies from a biocultural perspective. As anthropologists Michael Winkelman and John Baker explain,

> The biocultural perspective is based on the explicit recognition that humans are biological organisms whose primary means of adapting to the world is culture. Some of the most important . . . ways that humans now adapt to the world are made possible by the human propensity for religiosity. As a human universal, religiosity is rooted in our biology and given expression by our culture. . . . However religiosity arose, it now represents an evolved aspect of our biology that serves many adaptive [and maladaptive] functions. (2010, 28)

Religion, and our bodies, are both understood here through *naturalistic understandings*. This means that from this perspective our biophysical bodies have evolved; our propensity to believe in invisible agent/s, force/s, and/ or being/s and form systems of practice around these beliefs has evolved (Atran 2002; Boyer 2007; Taves 2011);[9] humans at the aggregate level have used and continue to use cultural mechanisms (including especially material aspects of culture) to shape ecosystems; and these cultural mechanisms have been and are still shaped by ecosystems. And per the anthropological record, until the rise of humanistic, Enlightenment, and now post-Darwinian thinking,[10] the key way all humans have ground cultural systems in a cosmology to make sense of both space and place and our roles within these has been through what we as scholars call "religion" (note this is an etic term germane to the academic discipline denoting the performed study of the same).

A materialist understanding of religion can supplement the biocultural view and both signal a needed return to the body in academia. The trench warfare of postmodernism as a rebuttal to the totalizing and essentializing meta-claims of modernism led many down a path of cultural relativity, word games, and solipsism. The grist of looking for power, and theorizing dynamics of power, made this rejection of modernist trajectories both needed and worthy. Yet the desire to do away with meta-claims and also the staying power of the very real forces of culture and biology, both, and a total lapse of taking larger bio-ecological context and trajectories seriously created a hole that a materialist turn to bodies, embodied intersubjectivities, and the more-than-human world is beginning to fill.

The scholar of globalization, transnational flows, and material religion, Manuel Vasquéz, helps add needed insights to this reclamation project of "recentering the body in religious studies" (2011, 21). His goal is to theorize a "fully somatocentric [bodily focused] non-reductive materialism" (2011, 44), against the to-date "somatophobic [bodily averse] internalism and ideal-ism" (2011, 59) that has largely defined religious studies. For Vasquéz, such a nonreductive materialism both takes seriously and is found at the confluence of biological (especially ecological and physiological) and sociocultural factors and it carries with it a "humbly agnostic" view of "'supernatural' sources of religion" (both 2011, 5). Rather, to make sense of practitioner's religious claims the somatocentric, bodily focused scholar looks to how people "build their identities, narratives, practices, and environments" (2011, 5). Therefore, a materialist explanation and exploration of religious phenomena must pay attention to how "neurocognitive networks and processes interact with social and cultural practices, which have a material density in and of themselves" (2011, 198). For Vasquéz people build religion with all the above, as it is all part of material reality within which we exist as biocultural beings. Our neural networks iteratively interact with, shape, and are shaped by social and cultural (material) practices, in bio-ecological spaces, to produce religion. Such bodily networks, mediated as they are by material cultural densities, are power-laden and power-shaped, while also being impacted by larger environmental trajectories. Thus for Vasquéz religion is materially enacted, embodied, and emplaced in emergent power-infused discursive-cognitive-biological-cultural communities of action. Together these categories combine to create a biocultural materialist understanding of religious bodies that sees, a la Vasquéz, "Religion and culture as embodied practices, artifacts, institutions, and environments [that] play a co-originating role [that] enable and constrain our being-in-the-world" (2011, 307).

Vasquéz's astute call for conceiving of religion as a bodily centered, enacted and emplaced material force in people's lives forms a key lens for my thinking through religious bodies in an era of rapid climate change. Here his discursive-cognitive-biological-cultural communities of action means that religious bodies are those shaped by religious discourses and conceptions at the meso-level, enable and constrained by being evolved animals in ecologies of place. Here the shaping is via neural pathways and the firing of synapses that take place during (religious) cognition. Religious bodies are therefore those, too, that reside within an even larger biocultural-based religious community. It is this material community that generates and maintains such discourses, and the bodily practices engendered by such discourses, as we will see for example later in the book with surfing. Here then religious bodies are also those shaped by performed religious actions that are informed by, and that inform, materialist and affective religio-discourses and religio-cognitive functions. This shaping occurs by engaging in performed bodily religious actions such as the following: ritual, music, dance, pilgrimage, ecstatic experiences, rites of passage, festival, fasting, bodily scarification and enhancement, devotional practices, taboo-based hunting, and many more. Religious bodies thus materially enact and embed meaning-making[11] performances and actions within discursive-cognitive-cultural-biological places, and the various communities that make these places. Here it can be helpful to understand religious bodies as performing a "religious dramaturgy." I use this term in the same way religious studies scholar Jill Stevenson grounds her research on evangelical Christians in the United States who practice what she terms "evangelical dramaturgy." For her "all dramaturgical systems . . . assume certain interpretations of representation, realism, enactment, spectatorship, and presence, in order to achieve particular aesthetic, ideological, and experiential effects" (2013, 4). The experiential, ideological, and aesthetic effects of religious dramaturgy are always performed by, in, and through religious bodies that dwell in ecologies of place, and are therefore always bioculturally performed within bio-ecological places.

It is important to say that the above is compared to other types of bodies with their own nonreligious dramaturgical systems: bodies and sports, bodies and politics, bodies and shopping, bodies and love making, bodies and medicine/healing, bodies and art/theatre, bodies and play, bodies and raising children, bodies and caring for elderly, and bodies and education (including in the academy).[12] Key in delineating bodily types is the insight from religious historian Rhys Isaac who writes that "meaning is so specific to context. . . . The whole categories of social-dramatic devices through which interaction communication—expression, direction, and ultimately coercion—may be accomplished is conveniently designated by the term *dramaturgy*. It will readily be seen that each culture and subculture has its own distinctive dramaturgical kit" (italics in original) (340, 350–351). The distinctive dramaturgical kit I will apply in case studies during the second half of the book is one grounded in part in a feminist new materialist, queer ecological epistemology to analyze the culture and subcultures of bodily practices in religious settings on a planet undergoing climate change.

To this nonreductive materialist dramaturgical lens can be added a new materialist feminist perspective. Such a perspective seeks to "move beyond discursive construction and grapple with materiality [by building] on rather than abandon[ing] the lessons learned in the linguistic turn[s]" (Alaimo and Hekman 2008, 6) of postmodernism and poststructuralism. Rather than inhabiting the space of somatophobic total language deconstruction and social construction created by these two turns, material feminism combines material with discursive lenses so that "nonhuman materialities [are] themselves [understood as] bona fide agents rather than as instrumentalities, techniques of power, recalcitrant objects, or social constructs" (Bennett 2010a, 47); that the world is populated with material entities where "the capacity of things—edibles, commodities, storms, metals—[are able] not only to impede or block the will and designs of humans but also to act as quasi agents or forces with trajectories, propensities, or tendencies of their own" (Bennett 2010b, viii). Here not only can forests think and speak since "meaning, broadly defined, is part and parcel of the living world beyond the human" (Kohn 2013, 20), but all material aspects of this living world have potential for agency.

Agency here can mean predetermined and premeditated actions by beings with nervous systems, with sense organs, and with evolved capacities to respond to stimuli around it. It can also mean the ability to impact, to affect, other beings and biogeochemical processes. So from this perspective, greenhouse gases affect other beings and cycles, giving them a form of material planetary agency. The feminist theorist Krizia Nardini explains that this emerging way of understanding reality, seen most clearly in new material feminism, recognizes that

> Ontology is conceptualized as a rhizomatic network of becoming(s) or material-discursive intra-actions. What we are witnessing is a reality with no copies and no originals, in which everything becomes intransitively, at various speeds and intensities, interconnected with other processes of transformation(s). Therefore, within this radical immanent ontology of becoming, there is no room for structural dualisms or representationalist instances. (2014, 21)

It is in everything become at-once with everything else, in the various interconnections and always ongoing entanglements where dualisms collapse, that agency is always present: present in those forces and factors that affect biogeochemical structures, but also in how these co-create material reality with all evolved beings.[13]

This multiplicity of materialist agencies helps create a trans-corporeal mesh of interacting matter within which our material bodies (and therefore religious bodies) are always embedded. Material feminism requires scholars to attend to "a creaturely immersion in the world" (Alaimo 2016, 146) and see religion as being a subset of "a multitude of naturecultures" (Alaimo 2016, 51) in a "fleshy posthumanist vulnerability that denies the possibility of any living creature existing in a state of separation from its environs" (Alaimo

2016, 167).[14] As bio-ecologies shift, all living creatures residing within will have to adapt. There is no separation.

The emphasis here is on any living creature, and assemblage of creatures, broadly (that includes at this level abiotic factors and forces)—all of it together co-creates the material world of which the human-animal (and all the other organisms within and upon our individual animal material bodies) is a derivative product and agent. Yet, human animals are agents who at the meso-level through culture and technology are able to impact the material reality of other creaturely agents. Since we are always embedded and impacted by the agency of material others, we cannot abstract ourselves out of this reality. Our theorizing must come home to our earthly biocultural context, and it must do so by investigating the always ongoing, multiple sited fleshy agentive dramaturgical interactions of religious bodies as sites of meaning-making and cultural construction.

Here material feminism is part of the ongoing "nonhuman turn [that] insists (to paraphrase Latour) that 'we have never been human' but that the human has always coevolved, coexisted, or collaborated with the nonhuman—and that the human is characterized precisely by this indistinction from the nonhuman" (Grusin 2015, ix–x). Material feminism posits that "current biophysical realities can only be approached through scientific captures of a multitude of intersecting biological and chemical, as well geological, transformations, which intermesh human and natural histories" (Alaimo 2016, 148). Our religious bodies and their dramaturgical performances have always been intermeshed with natural histories and earthly transformations; at times even overdetermining these histories and where coming transformations will be overdetermined by intersecting bio-ecological realities triggered by anthropogenic climate change. However, within this we must acknowledge as does religion scholar Beatrice Marovich that, "This posthuman, material feminist move beyond looks at both co-creation with, but also speaks to diffraction, disassociation, and distinction with fellow creaturely kin" (Marovich 2017, 323). In part II of this volume my focus will be on the former, with co-creation, including unintentional maladaptive co-creation, but hopefully with resilient and adaptive co-creation, too.

A posthuman, trans-corporeal, material feminist turn includes a "deep sense of skepticism about the power of knowledge and a high regard for the complexity and uncertainty inherent in the problems we face today" (Marocco 2008, 319).[15] Because of this, material feminism is an emancipatory approach in that it stands in solidarity with the following: (1) creaturely kinds that are victims of human material biocultures that are causing species extinctions and climate triage and (2) the poor and marginal in human societies who suffer the most under patriarchal empire. It invites us to rehabit our creaturely bodies (and requires us to see academia as constituted by such bodies in material places) in a way that is politically liberatory for all oppressed creaturely bodies, whether human or other kind. And for the immediate purposes of this book it is an emancipatory project to reconfigure "our spatial and temporal relations to the weather-world and [to cultivate] an imaginary where our bodies are makers, transfer points, and sensors of the 'climate change' from which we might otherwise feel too distant, or that may seem to us too abstract to get a bodily grip on" (Neimanis and Walker 2014, 559). Rather, our (religious) bodies are already enacted upon, implicated within and by, and drivers of material climate change—we must inhabit this dwelling place we have largely created through our use of fossil fuels and agriculture, and look soberly into the material future already here as an object larger than ourselves and of which we are already inside (Morton 2013). And it must get a grip on creating just biocultural places of solidarity, equity, and flourishing for all human body types. This is because in this bio-ecological, bodily, material dwelling, we must attend to power dynamics, arguing for a just transition to a post-carbon future. One where dispossessed and colonized bodies, of human and nonhuman kind, are allowed to thrive and flourish. This is the feminist praxis of liberation joined with new materialist insights. This merging holds promise for religious studies, but also the environmental humanities, and is one I invite other scholars to engage within their respective research and teaching.

The theorist who has probably done the most in helping usher in a liberatory critical posthuman turn in the humanities and that takes material agents and agencies seriously is Donna Haraway, Distinguished Professor Emerita in the History of Consciousness Department at the University of California, Santa Cruz. However, despite her history of theorizing past the human mind and body, Haraway has become critical of both posthumanism and techno-humanism, two fields of study she helped develop (2016). She maintains that both of these turns are stuck in the Anthropocene and are wedded to an outmoded belief in the progress of modernity. For Haraway we need to instead investigate "historically-situated relational worldings" (2016, 50) to reimagine, re-enliven, and reconnect as Earthbound beings with the rest of our fellow creatures on our shared planetary home. Haraway creates a new trope, the Chthulucene, to help make sense of the climate-changed world our actions have created. For her, the Chthulucene is mutually constructed not by autopoietic (self-organizing) systemic processes, but rather sympoietic (relations that are mutually beneficent between closely residing organisms) and symchthonic world-making where material realities mutually shape one another and per her use of the word symchthonic are "always partnered all the way down, with no starting and subsequently interacting 'units'" (2016, 33).

There are no autonomous agents in Haraway's reading of biocultural unfolding. Here she challenges the epistemological bedrocks of humanism,

Enlightenment thinking, modernity, and in many ways, the academy (see table 1 in the appendix for an overview of similar challenges). Rather, humans are instead part of the larger world of "abyssal and dreadful graspings, frayings, and weavings . . . in the generative recursions that make up living and dying" (2016, 33). For Haraway, the world is constituted by the always unfolding, ongoing interactions of "living beings' dynamic organizing processes [that are] looped, braided, outreaching, involuted, and sympoietic in terran worlding" (2016, 61). This is her key bio-ontological point, based upon her understanding of holobionts. Like holobionts, all forms of life, and therefore human animal bodies, are "symbolic assemblages, at whatever scale of space or time, which are more like knots of diverse intra-active relatings in dynamic complex systems, than like the entities of a biology made up of preexisting bounded units (genes, cells, organisms, etc.) in interactions that can only be conceived as competitive or cooperative . . . all of the players are symbionts to each other, in diverse kinds of relationalities and with varying degrees of openness to attachments and assemblages with other holobionts" (2016, 60).

Haraway's support of multiple-species becoming in a time of refuge and collapse is a normative project that brings her beyond posthumanism and into what she calls a "composist" (2016) understanding of bodies: we are all living, looping, braiding, co-creating, regenerating, consuming, and dying together in sticky, wet, moldy, creaturely becomings. To get our analytical hands dirty means that our theorizing needs to get to this level of depth. I invite our discipline, and the academy broadly, to start digging as time is of the essence. Each year we delay, including delaying our actions for liberation of oppressed bodies, brings us deeper into a CO2-braided, material co-becoming future that will be maladaptive to many fleshy bodies, including our own.

In my own composist digging I am most intrigued by Haraway's description of and theorizing through sympoietic terran worlding. The understanding that life co-creates through itself and its myriad creatures in the soils, seas, forests, mountains, and atmosphere radically decentralizes the human animal. Given bio-ecological rupture and biogeochemical shifts, this decentering must imbue our scholarship and theorizing with needed humility. Without such decentering, we will be unable to stand, swim, somersault, dance, crawl, and be present to creative projects of "partial healing, modest rehabilitation, and still possible resurgence in the hard times of the imperial Anthropocene and Capitalocene[16]" (2016, 71). The unfolding terran worlding of CO2, methane, water vapor, and radical sympoietic system shifts will impact, in ways currently undertheorized, "the global complex of animal religion" (Schaefer 2015, 15). This insight challenges us to generate a theorizing of religious bodies that recognizes we are assemblages (Schaefer 2015, 59) and have never been sovereign selves—which hopefully challenges us to reconceive of the academy through the same lens, leading to an adaptive praxis of education fit for bio-ecological assemblages and not sovereign agents. This task requires urgency, as all the bio-ecological assemblages we have created from the onset of the Agricultural Revolution 10,000 years ago are in sympoeitic refoldings and reworldings, unleashing powers and processes beyond the human. This book aids in this composist project.

BUT . . . WHAT ABOUT OTHER BODILY UNDERSTANDINGS?

My approach to investigating the human body in this book is built upon a certain techno-scientific discourse, one that as evidenced in this chapter is extremely helpful in understanding our bodies. However, in a book such as this it is important to recognize that other discourses and knowledge systems about bodies exist, including especially religious ones. Here I will briefly mention one: religious views of the body from Hinduism. There are many possible religious systems I could choose to offer as a foil to the naturalistic evolutionary view of human bodies offered earlier, including religious under- standings from biocultures that have viewed the world in ways material femi- nists argue. In choosing Hinduism I do so not to privilege Hinduism over any other possible religion or biocultures, or to perpetuate a history of Orientalist framing of bodies of "the other," but rather for three reasons: (1) one of my PhD qualifying exams was on religion and nature in Southeast Asia, with a heavy focus on Hinduism, so I have some fluidity in this body of literature; (2) I have some experience of Hinduism from traveling and doing research in India, and I regularly teach Hinduism in a variety of contexts so have a certain level of familiarity with Hinduism; and (3) climate change is already impact- ing religious bodily production in Southeast Asia, including at some of the region's most important sacred spaces. I will explore point three in chapter 5, so presenting some foundational knowledge on Hinduism here will help prepare the ground for those explorations. This said, I readily invite readers to actively search out other views of bodies besides that of Hinduism, or a techno-scientific evolutionary perspective, or to even go deeper into Hindu views of the human body than what I present below. Those readers famil- iar with Hinduism (especially readers who are experts in Hinduism and/or Southeast Asia) are invited to skim through or skip altogether this section.[17]

The colonial term "Hindu" is a placeholder for a wide variety of cultural stories, practices, ethics, dramaturgy, and texts that relate to a wide variety of deities, where this diversity has been building for thousands of years. In speak- ing of Hinduism and its conceptions of the physical body it is important to note there is no orthodox, normative view of the body that underlies it. Rather, Hinduism is multivalent, with many narrative myths, deities, conceptions of health and wellness, rituals, and other aspects we deem religious. These include the sacred dramaturgical importance of a wide variety of pilgrimage sites, the performance of religious dance and song, *puja* (ritual offerings to deities or holy persons), a variety of renunciate sects and practices, and the recognition of holy people and their *sidis* (spiritual power/s as a manifestation of *Shakti*, the divine feminine power), to name others. This heterogeneity has been captured in thousands of anthropological, archaeological, historical, phenomenological, and other studies. My effort here is to not distill these down, given this diversity of emic and etic scholarship and writings, as such an undertaking "would require a small library of volumes" (Rodrigues 2011, 2). Rather I give here a very cursory primer on key concepts of Hinduism as these relate to the physical body, bringing these into the larger conversation of this book.

To understand how the body is conceived, broadly speaking, in Hinduism we must begin with *śruti* literature. This is the literature that most Hindus believe religious seers, *ṛṣis*, perceived through communicating with the divine. As such, this body of literature is seen to derive from the holy/sacred/ divine itself, mediated through humans, and is authoritative in this context for most Hindus. The key body of *śruti* literature are the *Vedas*, which are comprised of four genres. It is in the *Vedas* that important origin myths are found, and thus are found early Hindu concepts of the body.

A foundational human origin story from this literature, found in the *Rig Veda*, is the "myth of the primeval man." This first human-like person was sacrificed by the gods and from its body parts came the four human *varṇas*, or classes, from which derived the Hindu caste system. From the mouth came the Brahmin, or priestly class; the arms, the *kṣatriya*, or ruling class; from the thighs, the *vaiśya*, or skilled workers and merchants; and from the feet the *śūdra*, or unskilled workers. Men from the first three classes participate in a coming-of-age rite of passage ceremony where they become "twice born," which enables them to enter the World of Fathers in the afterlife.

Related to this, and important in Vedic literature (2000BCE—1000BCE), is the concept of and concern for *ṛta*, or cosmic order. Maintaining cosmic order with the deities was the purview and function of the Brahmin religious leaders in the Vedic era. To this day this function of helping to maintain communal order and harmony with deities is one of import for the same hereditary lineage of priests. In early Vedic Hinduism this order was maintained through proper sacrifices to and at a fire altar. If the orthopraxy of these sacrifices was followed, then it was thought the community was living within accordance of cosmic order, or *dharma*. The priests who oversaw the sacrifices passed down the hymns, *mantras* (verbal chants and prayer formulas), and proper ritual protocol to their male lineage based on formulas and instructions that were originally shared to humans via the divine and recorded in the *Vedas*. It is this context that these texts are authoritative, for through their instructions those men who undertook proper ritual sacrifice and thus helped maintain cosmic order were believed to be reborn in the "World of Fathers." In this understanding of creation and social roles the twice-born Brahmins were, and still are, privileged, followed by twice-born men of the ruling class and skilled worker/merchant class.

Another key collection of writings are the *Laws of Manu*. These fall in the category of literature known as *smṛti*, or that which is remembered: sacred texts authored by humans and passed down as tradition, and not seen as being communicated to humans directly by the divine. Also in this category of literature are the Puranas, or origin myths and other stories associated with the variety of deities of the post-Vedic Hindu pantheon. The Laws of Manu contain community norms around which people are to live, proscribing as they do gender roles, marriage expectations, and a variety of other ritual and social rules and roles. Included in here, and which play a large role in Hindu conceptions of the body, are concerns with purity and pollution, or substances and rituals that purify the body and the soul, and those that bring pollution to the body and the soul. Related to these categories are astrological concerns with auspicious and inauspicious times and celestial events that impact the body (e.g., demons and bad spirits can come out during a solar eclipse and harm people).

Most Hindus also accept the doctrinal developments of the Upanishadic era of Hinduism. This era was a response to the sacrificial rites of the Vedic era and power of the Brahminic priesthood. The *Upanishads* are the final collection of *Vedas* and deal with various systematic questions about the nature of being, the nature of the self, and what happens after death. The religious, philosophical, and ethical answers provided still inform Hindu concepts of the body to this day, combining with insights of the body and social roles provided by the *Laws of Manu*. Of key import are a series of interlocking conceptual categories found in *śruti* literature, and upon which the *Upanishads* build.

The first of these is that Hinduism adopts a cyclical view of time, where all of creation goes through four eternally recurring epochs, or *yugas*. At the beginning of this recurring cycle the first *yuga* has total order and harmony. Each successive *yuga* loses a quarter of this so that by the fourth *yuga*, the Kali *yuga*, which many Hindus believe we are currently in, there is three-quarters vice, infidelity to customs and social mores, drinking, gambling, and other social ills. It is a time of degeneracy and signals that the god Vishnu will return and his return will cause the god Shiva to destroy the observable universe and all life forms, and the god Brahma will then re-create the universe and the cycle of *yugas* will repeat.

During the Upanishadic era religious teachers and philosophers began to formulate philosophical conceptions of the human soul. Teachings that form the core of Upanishadic literature and practice emerged amongst these *gurus*, renunciates, and holy men (and some women [Johnsen 1994]). Some of these conceptions provide a second set of interlocking categories that inform Hindu views of the self and body. It is important and necessary to report, though, that there were and are no monolithic answers to these questions about the nature of self and reality. A variety of philosophical, religious, and ethical answers emerged during the Upanishadic era and are to this day informative for different Hindus. Again, the reader is invited to engage such diversity if they so desire. However, one set of answers constellated into a level of explanatory power that does impact a vast number of Hindus, past and present. The first of these answers to the questions about the nature of reality and the self is that sentient beings in various rebirth realms contain an *atman*, or an eternal soul that is an aspect of the eternal divine essence of reality, *Brahman*. The rebirth realms include rebirth in various heavens or hells, or as animals, and of course rebirth as a human. These Upanishadic teachers maintained that where one is to be reborn is dependent upon their *karma*, or actions that bear fruit to lead to a more auspicious or inauspicious rebirth in the possible rebirth realms, depending on the nature of the action. The concept of *dharma* transitions during the Upanishadic era to refer to ones' duties in life as ascribed to them by their birth order, class, and gender. Fulfilling ones' *dharma* (with much of proper *dharmic* action outlined in the *Laws of Manu*) brings good karma and better rebirth. However, and key for Upanishadic Hinduism, the ritual power of the Brahmin priests and their entry into the World of Fathers via orthopraxy is directly challenged by the holy men (and women) of the Upanishads: broadly speaking, liberation is for all humans if they renounce, undertake yogic practices to purify the body and mind, cut through the illusion (*maya*) that humans are separate from the divine and in so doing escape the wheel of *samsāra* (eternal rebirth) and achieve liberation (*moksha*) from rebirth by achieving soul-union with the divine. The various yogic practices developed during the Upanishadic era function to cleanse one's soul of *karma* and help instead lead to liberation.

The above insights from the Upanishadic era on the nature of the soul, reality, and the body were a challenge to the Vedic sacrificial system and hegemony of the Brahmin class. However, such liberation typically was only for male renunciates. It also left aside conceptions of and interactions with the divine. Such conceptions of the latter are found largely in the *Puranas*, and it is the worship of the various Hindu deities (or the monist conception of one divine force of which deities are manifestations) that most Hindus dramaturgically practice today. The experience of *darśan* (sacred site with the divine who reside in *murtis* in purified temples [Eck 1993]), engaging in *puja*, including giving of incense and flowers (McHugh 2012), and imbibing sanctified and purifying food items (*prasād*)—these are the common religious bodily practices of most Hindus, regardless of class, gender, or location within the Hindu diaspora. It is believed that this love-based worship (*bhakti*) and relationship between a deity and devotee, akin to that between a loving parent and child, can lead to *moksha* in this life for all who worship the divine. Undertaking pilgrimage, especially to Varanasi (Justice 1997), as well as engaging in other acts of worship on auspicious days and engaging in purifying behaviors also aid the devotee on the path of building good *karma* and achieving liberation.

There is one last category in the Hindu worldview that relates to concepts of the body, specifically, and does so in a way that ties the above categories together. This is the concept of *gunas* as articulated in the *Laws of Manu*. As discussed, karmic actions determine where one is reborn on the wheel of *saṃsāra*. One's station on this wheel is seen to be hierarchical (Nelson 2006, 185) and is a result of one of three possible psychological dispositions, or *gunas*. These are the "'qualities' or 'strands' that [combine to make] up the entire range of phenomena, mental as well as physical" (Nelson 2006, 185). The three *gunas* are hierarchical and are (1) *sattva*, or the psychological and thus physical disposition of goodness and lucidity, which correlates to "intelligence, creativity, and spirituality" (Nelson 2006, 185); (2) *rajas*, or "energy," which is the psychological and thus physical disposition of "passion and dynamism" (Nelson 2006, 185); and (3) *tamas*, or "darkness," the psychological and thus physical disposition for "ignorance and lethargy," which in the Laws of Manu is associated with "confusion, sensuality, inability to reason, lack of intelligence, greed, sleepiness, incontinence, cruelty, atheism, and carelessness" (Nelson 2006, 185). *Sattva* is a psychological and physical attribute attributed to health, spiritual advancement, and the divine, *rajas* with humans, broadly, and *tamas* with animals. Therefore to understand most Hindu conceptions of the body, one must understand the hierarchy of these psychological and thus physical attributes.

In sum, most Hindus understand physical bodies to reflect karmic action from prior rebirth, where following ones' *dharmic* duties influences future rebirths. Social roles are still largely stratified, gendered, and hierarchical, dating back to expectations articulated in the *Laws of Manu* and the myth of the primeval man. If we are male or female, or a non-binary gender; if we are a first-born son or last-born daughter; if we are born with physical disabilities or with pure health; our education and job expectations—all of these relate to *karma* and *dharma*. The best path for cleansing of *karma* is to participate in purifying rituals and in interacting with purifying materials (especially products from cows), and to also engage in practices of *bhakti* at home and temple shrines, practicing devotion to the divine. The psychological dispositions and therefore physical attributes of people and their bodies are a direct result of the *gunas*, and the overall goal is to leave the body by achieving liberation either through yogic practices or more specifically engaging in various religious rituals and pilgrimages and engaging in devotional practices to a deity.

This will allow the *atman* to rejoin the divine and not be reborn in a physical body in any possible rebirth realm.

Given the above as an explanation of visible reality and all bodies therein, it is the case that some Hindus claim that this understanding of reality (cosmogenesis and cosmology) reflects an early understanding of a Darwinian worldview based on evolution. From this emic perspective Hindu views, broadly speaking, of *yugas* and specifically the many *avatars* (physical manifestations) of the deity Vishnu are a proxy for an evolutionary worldview. From a scholarly point of view this is a false claim and should be read as a *post facto* attempt by Hindu apologists to parallel and make agree Hindu cosmology with contemporary evolutionary science. The understanding of Darwinian evolution and modern science, broadly speaking, rests on very different epistemological and ontological foundations than do Hindu religious views that date back to the Vedic era. This is not to say the Hindu views shared above are "false," but it does mean it is inaccurate to claim they are a reflection of an evolutionary worldview that is both outgrowth of and provides the foundation to modern science.

The reader should be reminded that in the above evolutionary understanding of the body, and then Hindu understanding of the body, I have spoken in broad brush strokes and did not enter into the myriad internal-debates, past and present, over conceptions of the body within both discourses. Rather, my goal here has been to recognize heterogeneity in conceiving of human (religious) bodies. My hope, though, is that such a move helps to elucidate why an evolutionary perspective is valuable in helping to analyze changing experiences of religious bodies in the Anthropocene. A key insight is that an "anthropology of the senses" (Seremetakis 1994, 4) that takes "discussions of the sensuous body" (Stoller 1997, xv) seriously must emplace our sense-bound biophysical creaturely bodies within a planet undergoing climatic shifts. These shifts will impact our bodies' responses, and thus the responses of cultural-religious bodies, to the shifts, as explored in later chapters. We must ground such an academic understanding of these shifts within the baseline of evolution and bio-ecological understandings, or else we are perpetuating severely blinkered scholarship insufficient to the task of helping to create resilient practices of dwelling on a much warmer planet.

RELIGION, BODILY DRAMATURGY, AND CREATURELY COMMUNITY

Given the above discussion I want to revisit the operational definition of religion I adopt for this book.[18] Three further points about it are needed: (1) congruent with a biocultural analytic, the cosmological/mythical narrative of a religion and the ethics, practices, teachings, rituals, doctrines, and identities derived from and related to it are shaped by and shape human/nature interactions; (2) religious systems are not and have never been static; and (3) key, these religious systems are embodied and emplaced in material practices and bodily (singular and communal) movements. Phrased differently: there is no religion without the more-than-human bio-ecological world, and there is no religion without bodily practices.

This insight is especially pertinent because in the current understanding of human/nature interactions

> genes and DNA do not account for everything any more than the old drives and instincts or the older gods and spirits did. Environment and natural selection are today's other costar actors. Together, DNA, environment, and natural selection make us who and what we are. And we are embodied, be-brained, mammalian, bipedal homonoids. Some of us ritualize and some do not. So the question is not "Do we ritualize?" but "Should we?" The ritual question is morally and practically driven. The long form of the current existential ritual question, then, is: Should humans ritualize in order to be attuned to nature and thereby avoid planetary destruction? (Grimes 2005, X)

In other words, ritual-in-community (a form of religious dramaturgy) guided by cosmological stories help embed and emplace us in a biocultural place; they contribute to our dwelling. As the anthropologist Tim Ingold explains, human animals are always already "in the context of an active engagement with the constituents of [their] surroundings. I call this the 'dwelling perspective.' Humans . . . are brought into existence as organism-persons within a world that is inhabited by beings of manifold kinds, both human and non-human. Therefore relations among humans, which we are accustomed to calling 'social', are but a sub-set of ecological relations" (Ingold 2000, 5). Here we see again that we are always already in sympoeitic material relations of mutual becoming with the natural, ecological world of which we are part and which we are entwined, embedded, and both shape and are shaped by. We cannot escape our animality, and the biophysical reality that we are always first and foremost (and have never not been) social primates embedded in ongoing, unfolding, chthonic, material bio-ecological relations. The sooner our discipline, and the academy at large, moves to this place of scholarly dwelling, the sooner we can aid in generating resilient adaptive creaturely habits and actions. And for religious studies (and theology) scholars, the sooner we can also ask pertinent questions about how religious dramaturgies may and can be either adaptive or maladaptive given our bio-ecological species dwelling.

Ingold also provides a rich foundation from which to help conceive of (religious) humans as embedded biophysical beings in mutually becoming

material ecoscapes. He points out, akin to material feminists, that our locative habitats and embodied actions occur in continually co-created places.
For Ingold we need to adopt a dwelling perspective to make sense of how
our understanding of self and human culture does not begin within or originate prima facie as a building block emerging from our abstracted-out-of-
nature Archimedean and hermetically sealed human consciousness. Rather,
we need to theorize bodies and culture, both, from the standpoint of "the
immersion of the organism-person in an environment as an inescapable
condition of existence. From this perspective, the world continually comes
into being around the inhabitant, and its manifold constituents take on significance through their incorporation into a regular pattern of life activity"
(Ingold 2000, 153).

For this book, however, it is as important to recognize that the bodily,
fleshed, sensory-somatic person-inhabitant continually comes into being as
an organism in biocultural worldings. For many this bio-ecological bodily
becoming includes religious dramaturgical practices and lifeways and patterns of life activity. And in seeing religious production via bodies as a key
subset of culture, we must understand that "the differences we call cultural
are themselves biological . . . form, I argue, is not received by the organism-
to-be at the point of conception, but generated within the dynamic functioning of developmental systems. And through contributing to the environmental
conditions of development for successor generations, organisms—including
human beings—actively participate in their own evolution. There can, then,
be no specification of the essential form of humanity independent of the relational contexts in which human beings *become*" (italics in original) (Ingold
2000, 292). Or as Ingold writes in another context, "all creatures, human and
nonhuman, are fellow passengers in the one world in which they all live,
and through their activities continually create the conditions for each other's
existence" (2005, 503).

This means that the material, relational becoming of dramaturgical religious bodily practice occurs with other creatures and their dwelling and
agency, as this creates the conditions for our dwelling and agency. Even
the fossilized remains of agental-becoming of past creatures now literally
fuels the heating of our earth—more agency, and thus more material effects
and impacts on our material bodies! It is hoped that by briefly exploring the above basic evolutionary understanding of our bodies (cognitive,
emotional, physical), and then bringing to this material feminist lenses, a
context was created that allows for me to theorize the performance of religious bodily responses in ecologies of place to climate change. Without this
needed background it becomes very hard to analyze how these embodied
and performed responses may influence religion and religious production in
the decades to come.

NOTES

1. In this chapter I focus on the cultural development of anatomically modern human animals. This does not mean that other primates, past or present, did not have culture. Or that other mammals do not have culture. Or nonmammalian species do not have culture. The respective forms of culture of these planetary others would have, and continue to have, impact on their evolutionary development just as much as human culture impacts ours. And our cultural practices influence the evolutionary (and cultural) development of these planetary others, and vice versa. Because culture is material and physical, as well as being semiotic, it is part of evolution and as such predates and therefore influences the always ongoing evolutionary development of *Homo sapiens*. Consistent with this framing, though, is that symbolic communication and culture predate the emergence of *Homo sapiens*. Other animal kinds have forms of communication, including even symbolic, and other creature kinds of many species type have culture. Therefore culture and communication are part of the bio-ecologies within which human animals, and human culture, have bioculturally evolved. It also suggests that material practices were passed down haptically, by touch and observation, while language was also developing, such that shared meaning is as much bio-material as it is verbal. I thank Richard Carp for these insights.

2. "Climate changes" is a term utilized by glaciologist and National Geographic Explorer M Jackson. She visited the College of Charleston in spring 2018 and explained she uses this term as it better captures the stochastic, nonlinear, and place-specific impacts of climate change: these changes are always localized, leading to a variety of climate experiences unique to communities-in-place. Thus, climate changes.

3. See here a statement from the editorial staff of *Science Magazine* in response to the George Floyd murder and Black Lives Matter protests of May and June, 2020. While a laudable statement, it leaves out disproportionate power differentials along lines of gender, as well, and does not speak to how science helps drive the violent dispossession and extraction of "resources." On this I again point the reader to Kathryn Yusoff's (2018) brilliant and powerful *A Billion Black Anthropocenes or None*.

4. The "model of culture" for which Slingerland argues is one that sees social instincts as "metaphoric extensions of 'ancient ones,'" never far removed and always less vivid" (2008, 214). This understanding challenges the dangers of taking social constructionism to its logical extreme, and instead accepts that there is an observable, objective biophysical reality that exists outside of the conditioned mental/emotional/cultural framing we bring to it. For more on this, see the 2018 *The Atlantic* article "What an Audacious Hoax Reveals About Academia," by Yascha Mounk; and Barbara Epstein's "Postmodernism and the Left" (1997) on the 1996 "Sokal affair." I thank my religion and nature colleague Bernard Zaleha for sharing these sources. I will briefly revisit the topic of social constructionism and the category of "nature" at the beginning of chapter 3.

5. "Symbiogenesis, literally 'becoming by living together, refers to the crucial role of symbiosis in major evolutionary innovations" (Aanen and Eggleton 2017, 99).

6. This principle informs sustainability discourses, where we are urged to act with caution (precaution) when bringing new technologies and new behaviors into ecosystems. In principle all new interventions should be undertaken with severe, monitored caution until we have a better understanding of how these new behaviors will impact the ability of life (and thus humans) to sustain itself in ways that enable flourishing, thriving, and health. Champions of the precautionary principle argue it should guide our science, politics, economics, and education. For example, before unleashing the internal combustion engine upon the planet, we should have with precaution seen how cars, fueled by such engines and the fossil fuels that power them, would impact a community: how the people therein travel, live, behave, and interact. And how the material impacts of cars on the local bio-ecology impacts overall bio-ecological community health. With this data in hand, an informed choice could have been made on whether to continue with using fossil-fuel based engines and cars at planetary scales. The precautionary principle would hold that when measured against community health that the benefits of automobiles are non-existent, especially when compared against impacts such as road kill, vehicular accidents, fracturing of the social contract, and climate change, and that we should therefore have found other ways to travel and organize transit in and between communities.

7. Including our internal environments—there are countless organisms within our bodies, especially our intestinal tracks, that have coevolved with us and whose sympoietic functioning keeps us alive!

8. On these terms, see my discussion of Donna Haraway's work in the section "Theorizing Bodies and the Environment, or Material Bodies in the Academy, Redux."

9. For helpful critiques of a reductive, computational cognitive understanding of religion, see Vasquéz (2011, chapter 7).

10. Although there are now post-supernaturalistic, post-Darwinian religious systems (Taylor 2010; Gould 2005; Kleiner 2003); as well as religious systems that are perspectival based and may entirely eschew any concept of the supernatural, adopting instead a "multinaturalist" view (Viveiros De Castro 2015, 196), which includes interacting with and experiencing material bodies, presences, and features of bio-ecological places as part of religious understandings and dramaturgies.

11. And to be clear: the intended (or unintended) meaning may be contested, ambiguous, unclear, challenged, and/or changed by those involved in making and/or observing it.

12. Although all of these activities can occur in religious settings.

13. I struggle here, as do most scholars working in new materialist spaces, to capture this in words. For example, the material ecocritics Serenella Iovino and Serpil Oppermann share key questions that scholars have to address in these new materialist spaces: "how do we define the field of our experience of material natures? And, secondly, how do we correlate discursive practices (in the form of political categories, socio-linguistic constructions, cultural representations, etc.) with the materiality of ecological relationships? On what ground is it possible to connect these two levels—the material and the discursive—in a non-dualistic system of thought?" (2012, 76). I have not myself worked through answers or strategies to address these questions.

Rather, this book is my own initial efforts at so doing, and I look forward to using new materialist lenses in continuing to theorize what agency is, who has it, how, and the implications of this, in future articles and books.

14. Similar insights and cultural understandings have long been present in indigenous religions. The material feminist move can be seen as a development that parallels such long-standing biocultural, traditional ecological knowledge dramaturgies, but that does not occupy similar ontological origins.

15. Here, too, I myself must be leery and cautious of a reductionist scientific fundamentalism—material reality can be approached, accessed, and understood via different biocultural modalities than just the natural sciences. However, the understanding of biophysical evolution, and also climate change, provided by the natural sciences is to-date the most accurate way of understanding both processes: we know the material reality we are part of and engage is here due to evolutionary processes, and that anthropogenic climate change is caused by greenhouse gases trapping sunlight energy.

16. The era marked by European and then global enclosures, powered by fossil fuels and industrial and post-industrial technologies, that set the stage for today's global corporate capitalism. This era is co-terminous with the onset of the Anthropocene, and some scholars prefer the phrase Capitalocene instead, as it is not the human animal per se that is causing biogeochemical shifts at planetary scales, but the specific capitalist economy that is driving such behavior. For more, see http:// capitalocene.org/ and its location of "the climate crisis in the historical context of colonialism, orientalism and extractivism." It must be noted that the modern academy is both offshoot/manifestation of the Capitalocene and helps to perpetuate the Capitalocene (accessed May 25, 2020).

17. For those readers interested in going beyond the textbook "Hinduism 101" presented in this section I share here a preliminary list of suggested scholars based in the US who work on Hinduism: on religion and nature see David Haberman, Whitney Sanford, and Christopher Key Chapple; on the politics of representation of Hinduism (and Islam), especially in colonial contexts see the work of Audrey Truschke and Robert Yelle; on goddess worship and mysticism in Hinduism see the work of June McDaniel; on Hindu sacred texts see the work of Anantanand Rambachan; on Hindu theology see Jonathan Edelmann; and on Hinduism, broadly, see the work of Wendy Doniger, Martha Nussbaum, Bruce Sullivan, Diana Eck, and Vasudha Narayanan.

18. I paste it again here: religion (and thus the embodied and emplaced human-produced dramaturgical systems of) consists of a cosmologically/mythically grounded understanding of humans in community in relation to agent/s--power/s--being/s believed or experienced to have communicated (and often to continue to communicate) with humans-in-community guidelines for living that relate to and are centered upon interactions with the agent/s--power/s--being/s in question. Such agent/s--power/s--being/s may be visible or invisible, active in the past or currently active, and on-planet or off-planet (or both). The communicated guidelines shape and create systems of ethics, doctrines of belief, and bodily dramaturgical practices that inform an understanding of self and of collective community.

Chapter 3

Cultural Narratives and Science

I begin this chapter's journey by reflecting on a key insight shared by the same feminist philosopher of science, Kate Soper, whom I quoted last chapter on meso-level adaptation. This time her astute insight addresses the role of language, power, and politics in speaking of categories such as nature and the body, both, and doing so in the cultural trajectories of social constructionism and scientism. In other words, what do we mean when we speak about "nature," when as some argue categories of speech simply reflect social constructions? She points out that in English the term "nature" carries three possible meanings (1995):

1) nature in a metaphysical sense: here "nature is the concept referring to the difference and specificity of humanity" (1995, 319). By this nature can mean either metaphysical naturalism, where human animals have a sameness, typically based on an evolutionary understanding of life, with the rest of the observable world and universe; or a metaphysical anti-naturalism, where humans are different than other animals and thus are inexplicable in terms of the latter. In this view nature is contrasted with humanity and this dualism is policed, including in academia. I stand firmly in the former view of metaphysical naturalism, and argue that the policing of this fallacious dualism has blinkered our understanding of religious bodies as they are already responding to climate change.

2) nature in a realist sense—nature in this semiotic setting refers to structures, processes, and causal powers that operate in the physical world and that the natural sciences study. Here human animals are understood as being subject to laws of nature, which include causal forces that include human beings. This work sits firmly in this understanding of nature, as well. This is most obvious in my use of climate science, and also evolutionary sciences, to help think through the nature in a realist context of our religious bodies.[1]

3) nature in a lay/surface sense—nature in this semiotic context refers to ordinarily observable features of the natural world. Nature in this setting is a term used in everyday speech that reflects an experienced set of phenomena: "Oh, it's their nature to act that way so no worries"; "That nature hike through the ravine was beautiful."

The philosopher Ted Toadvine utilizes an ecophenomenologist perspective and shares similar insights as Soper to help think through the implications of both nature in a metaphysical sense and nature in a realist sense. He writes that "The 'environment [i.e., nature],' as a reification that stands over against the human subject, implies an artificial division between nature and humans and encourages us to view nature as a collection of things rather than in terms of mutually constitutive relationships" (2009, 6). The guiding analytical lens of this book is precisely this: humans are created by, and thus co-create, mutually constitutive relationships with the rest of the natural world. This is how life on our home planet unfolds beginning with the formation of cellular life, and will continue to unfold. However, one of the biggest changes we are mutually co-creating through our use of ancient sunlight to power industrial lifeways is climate destabilization: climate changes are here. They are real. And they may impact our species in a realist way that severely diminishes our bio-ecological flourishing. And such nature in a realist sense will continue co-creating life long after humans are no longer an extant species on our home planet, whether we make it through climate change or not.

Thoroughly related is that we are derivative products of sunlight energy that is cascading through trophic flows, both terrestrial and aquatic. These flows literally provide the evolutionary context for our bodies existing and reproducing, and thus form the foundation upon which culture resides. The sympoietic biotic (and, thus, abiotic) aspects of what constitutes nature in a realist sense has its own agency (I write this in September 2018 as Hurricane Florence is acting upon the US Southeast), and is filled with other bodies that have agency. There are biogeochemical processes in nature in a realist sense that shape the bio-ecosystems within which humans make culture and perform religion via our animal bodies. Any theory of bodies that does not take these two conceptions of nature as shared by Soper seriously is not complete scholarship. This, all of this, is the context of religious dramaturgy, past, present, and into a climate-changed future.

However, I am challenged in writing this chapter, and this book, broadly, by insights from queer theory (QT). To begin, QT "does not merely criticize the centrality of certain categories, [but also] challenges the very idea of categories" (Bauman and Stenmark 2018, 8). QT posits fluid identities and porous borders, and even questions the reality of borders: there is and never has been an archetypal species, or ecosystem type, or phenotype. Rather, nature in a realist sense as seen from QT is an intermixing dance of co-constructive, mingling, sexing, dancing cells, genes, organisms, weather patterns, chemicals, and atoms; that at its core nature in a realist sense as a "whole system of identity is unstable because gender is unstable . . . reproduction and biological sex are . . . unstable and constructed" (Bauman and Stenmark 2018, 11). In other words, if there is no stability at the level of sex and gender, then there's no stability at the system of organismic identity, and thus no stability at larger scales. This is not to be interpreted that tactile, bordered entities do not exist—clearly there is a Todd LeVasseur writing this book. I have a body that is largely discrete, and will be bounded by my skin for about 75 years, give or take, after which the elements that make this body will become something else (Todd LeVasseur's temporary boundaries will break down). Yet my body contains holobionts. There are as many bacteria in my body as cells. And my skin—it can grow cancer because of too much sun, and gets itchy from poison ivy. And my cells will not replicate if I do not imbibe water and calories. And if the ambient temperature gets too hot or cold my organs may shut down. So QT proponents do not discredit the natural sciences, or the insights of, nor do they say there aren't bodies that have agency and autonomy—they do question the heteronormative assumptions they believe undergird the language of and categories posited in the natural sciences, especially when it comes to conceiving of species.[2] This critique extends to assumptions that bodies in a realist sense are autonomous and thoroughly individual. And it extends to assumptions that ecosystems, or for this book, bio-ecological places, are autonomous and thoroughly individual. Such assumptions of autonomous individuality form the bedrock of (Western) epistemologies and ontologies (including in many religious ones), but as the disparate literature I am bringing to this book evidences, these assumptions are canards. They do not hold up to nature in a realist sense and the intermingled material and biogeochemical flows that sympoietically shape the bio-ecologies of our planet within which we always dwell. Nor do they hold up to the climate changes already in play, and that through co-evolving and interacting tipping points will only become more maladaptive to our bodies in the decades to come and beyond.

Which brings me to a challenge I have as an author who is engaging QT as a new academic home, namely, that sometimes metanarratives are reality—there is danger in "queering" climate change, for example. We really do have an atmosphere in a realist sense with over 415 ppm of CO_2 that is heading to 500 ppm CO_2 and beyond, and this is very likely going to lead to catastrophic die-off or outright extinction for many extant species types, followed by a new flourishing of future types of intermeshed bodies and agencies. I do not wish to "queer" this reality: I am fine with a category of pre-350 ppm CO_2 atmosphere and planet that has ice caps, and a category of a post-350 ppm CO_2 atmosphere and a planet losing ice caps and biodiversity. I do wish, however, to "queer" in subsequent chapters the lived realities of religious bodies as these experience a climate-changed world. The hermetically sealed category of "human" that undertakes religious dramaturgy such as rituals and dance and other bodily practices in religious contexts to interact with the sacred, with the natural world as mere backdrop, is a false and theoretically bankrupt category. Bio-ecological, biogeochemical, and evolutionary material realities lay bare such bankruptcy. This book joins with QT, material feminist, critical race, Indigenous Knowledges (IK), and posthuman scholarship to deconstruct such a fallacy of misplaced concreteness that privileges the human as separate/different from/superior to the nonhuman, including within religious contexts. And within this, privileges Cishetero White Male bodies.

So while I see QT as a needed liberatory movement, I want to recognize that some may interpret QT's pushback on binary categories as a reason to not accept the reality of climate changes. QT is not arguing against climate changes, just as William Cronon was not arguing there is no nonhuman nature that really does exist in the concept of "wilderness" (1996). His essay was clear about this, although it was used to say if wilderness was a made-up social construct, then there was nothing "there" to worry about preserving or protecting. Of course Cronon did not make this argument! And while QT may say that the Holocene was actually created by mixing microclimates at various scales, the reality is that the Holocene weather patterns which our agriculture and civilizations have evolved within are as a whole rapidly shifting and such shifting is maladaptive to ours and many other species.

While the above caution is important, this work stands in solidarity with the efforts of QT, and especially the queering of religion and nature theory. I celebrate in agreement the insights from Daniel Spencer who writes that "queer(y)ing opens up new and transformative possibilities, so critical in the perilous times of the Anthropocene" (2018, 18) and that by engaging in queer(y)ing we can all co-create a "dynamic *restorative* paradigm that honors human and more-than-human agencies" (2018, 18). The need for this in academia cannot be understated. I stand too in solidarity with Whitney Bauman's queering of climate change "in which agency and causality is distributed among and between multiple planetary actants" (2018, 112).

Queering religious bodies in an era of rapid climate changes means we need to back away from abstractions and reifications, where we assume our categories are free floating and account for actual reality. There is no religion, body, performance, economy, globalization, nature, weather, or any other fixed essence of, that elides and covers up the "symbiotic real" (Morton 2017, 60–63). Clinging to these erroneous categories leads to an estrangement and alienation from co-created reality; an enstrangement from the understanding that all exists because of "the sensuousness and specificity of nonhumans and the sensuousness and specificity of creativity" (Morton 2017, 60). Achieving this realization for Morton is the foundation of what he calls ecological thought, understanding that all is interconnected (Morton 2010); and ecological awareness: "coexisting, in thought and in practice, with the ghostly host of nonhumans. Thinking, itself, is one modality of the convocation of specters in the symbiotic real" (Morton 2017, 63). Scholarship of religious studies, and of religious bodies, must proactively address this in its theorizing, especially to better capture biocultural dramaturgies in an era of global warming.

Such ecological thought and awareness brought to scholarship should result in new and as yet unanticipated developments, as well as the nonhuman turn. It should also join with material feminism and the questioning of representationalism and hegemony of language and instead signal a shift to performativity: "matters of practices/doings/actions [and] questions of ontology, materiality, and agency" (Barad 2008, 122). What do humans practice and do with religion in the context of materiality (ecosystems and ecologies of place) where the agency of the nonhuman is pushing back in ways that call for "earthly multispecies recuperation and resurgence" (Haraway 2016, 8), if we could only stop to listen to and interact with such agental forces? For example, while studies of religion and healing across various traditions recognize the agency of spirits or black magic or *karmic* actions as influencing illness, disease vectors will increase in geographic range and number of human bodies impacted in the coming decades. This is because climate changes will impact the habitat range and reproduction rates of such vectors, exposing them to new human hosts. Theories and studies on healing and the body will need to account for this new phenomenon, while hopefully contributing to advocacy for the management and treatment from various cultural healing modalities of such vectors so as to minimize human (and nonhuman) suffering. Scholars, regardless of discipline (another reification that needs to be queered) must begin to ask such questions, especially if we are to navigate the coming political, economic, and material "global landscape of Anthropocene inequality" (Purdy 2015, 6).

HYBRIDITY

The above is not to say that nurture is not as important as nature, but that any nurturing will already be done to bodies with genetic predispositions at both species and individual levels; as Paul Ehrlich explains, "*Every* attribute of *every* organism is, of course, the product of an interaction between its genetic endowment and its environment" (2000, 6). Or as political scientist and sustainability expert Leslie Thiele writes, based on his exploration of evolution and environmental ethics, "genetic leashes" exist, but their length and use can be impacted. For humans, and therefore religion and bodies in a time of climate chaos, the two key meso-level determinants are the "cultural field of play" within which we live that contains our respective social mores; and our brains, given that neural information impacts us via cultural upbringing more than genes (Thiele 1999). Thus the nurture of both nature, but also culture, influences our capacities for belief and behavior, both. Here our understanding of modern categories related to nature/culture and human/nonhuman (Latour 1993, 10–11) continue to inhibit our understanding of ourselves as biocultural beings, and therefore how religion is a biocultural phenomena based upon and contributing to morphisms (Latour 1993, 137) and where our bodies are constant sites of encounters (Agard-Jones 2013). In as much as religion is a dramaturgical skill developed and used to make sense of biocultural places, it would help our theorizing to see it as such. From this perspective the industrial revolution, and thus post-industrial and post-materialist religious dramaturgies, have replaced more localized dramaturgical artifacts of religious bodies in action. As anthropologist Tim Ingold explains hunter gatherers who did not have their perception of and relation with bio-ecologies severed by machines maintain

> possession of acutely sensitive skills of perception and action. Yet as properties of persons, developed in the contexts of their engagement with other persons or person-like agencies in the environment, technical skills are themselves constituted within the matrix of social relations. Hence, insofar as they involve the use of tools, these must be understood as links in chains of personal rather than mechanical causation, serving to draw components of the environment into the sphere of social relations rather than to emancipate human society from the constraints of nature. Their purpose, in short, is not to control but to reveal . . . artifacts, too, may be grown, and that in this sense they are not so very different from living organisms. . . . Just as the form of the organism is not prefigured genetically but arises through a process of growth within a morphogenic field, so the form of the artefact is not prefigured culturally but arises through the unfolding of a field of forces that cuts across its developing interface with the environment. (2000, 289–290)

This suggests that to better understand the artifact of religious dramaturgy performed by biocultural bodies in bio-ecological places, we need to see how these dramaturgies arise and unfold in the dwelling places of bio-ecologies and the social relations such places support. Because we are always emplaced biocultural beings, our dramaturgies are learned and their performance informed by perception and conduct in material bio-ecological places. As these places shift, so will our social relations and our religious dramaturgies, as these encounter and are shaped by larger morphogenic fields.

For Bruno Latour such morphic encounters can include technomorphisms, zoomorphisms, phusimorphisms, ideomorphisms, theomorphisms, socio-morphisms, psychomorphisms, and how these interpenetrate and interrelate (1993, 137). Given such hybrid embodiments and always-ongoing morphic mixing, "the body is only interpretable within a specific culture and context" (Richardson 2014, 3), where that context has always been shaped by ecoscape flows. Whether these ecoscape flows have been and still are germs (Diamond 1997), or ice, dirt, and worms (Worster 2015), or soil carbon (Sandermana et al. 2017) the point remains: we are derivative organisms that live within evolutionary parameters and ecosystem flows. And like all organisms in bio-ecological places, we can and do impact these flows. Whether this impact is cellular, as seen in persistent organic pollutants passing up the food chain and now entering breast milk; or as melting our glaciers and ice caps due to use of fossil fuels, the results always end up impacting our cultural production, including of bodily religious dramaturgies in bio-ecological places. These impacts will increasingly influence and determine our ability to function as resilient biocultural animals. This is because "the forms of [religious drama-turgical] artifacts, like those of organisms, arise through processes of growth within fields of relationships" (Ingold 2000, 291).

AGENCY, BODIES, AND PERFORMATIVITY (OF CULTURE-BOUND BODIES)

In his study of "Gain pilgrims" (2003, 12) and their perceptions of and interactions with sacred geographies, the environmental humanist Adrian Ivakhiv explains that to properly understand how such pilgrims conceive of sacred landscapes we need to change our lexicon. Rather than seeing the creation of sacred places by New Agers as being a social construction where these and other neo-pagans project their views of a sacred nature onto a pas-sive landscape, we must use terms like "enactment, enrolment, enlistment, performance, dialogue, network-building, choreography and orchestration" (2003, 12) to more accurately theorize the phenomenon he encountered. This is consistent with the religious dramaturgy lens I utilize through this book, where our bodies performatively act in ways religious in ecologies of place. But key here is that in these places "the nonhuman should be considered an active agent, and in fact more than a single agent, within the analysis of sacred geographies. Nonhumans may not be agents in precisely the same way that humans are . . . . But they participate in the networks forged between human social groups, cultural discourses, technological systems and ecologi-cal relations" (2003, 12). Ivakhiv's fieldwork witnessed the encounter of New Agers and neo-pagans with sacred landscapes in Sedona, Arizona (USA) and

Glastonbury, England, leading him to conclude that "Rather than being simply imposed onto pre-existing landscapes, religious meanings may, in part,
emerge reciprocally with landscapes, cultures and practices, involving not
only human agents but other forms of agency as well." (2003, 24).

Recognizing agency of nonhuman others, whether animate (organisms) or
inanimate (weather patterns, sequestration of carbon in coal or shells), places
this book in dialogue with the new animism of anthropology and indigenous
studies. This movement builds on the category of animism, but recognizes the
pejorative and colonial baggage that this term carries. New animists recognize
a key insight of "animistic worldviews . . . that the plant, animal, and human
realms interpenetrate" (Hall 2011, 100), where such interpenetration brings
with it kinship relationships and therefore "obligations of responsibility, solidarity, and care" (ibid, 100). These obligations form part of a sacred ecology,
itself grounded in a sensibility that "the world [is] constituted by powers that
[take] the form of Persons" (Harrod 2000, xiv). Such a worldview merges
religion, the environment, and morality through rituals, ceremonies, sacred
stories, and dreams to build a symbolic understanding of a sacred ecology
that is built upon reciprocal relations between kin networks in a holistic context where "the land is sacred" (Martin 2001, 12). In chapter 7 I will explore
emerging dramaturgies of post-materialist, Western sacred ecologies that are
tethered to similar understandings of sacrality: that the land is alive, filled
with powers, has agency, and can be interacted with in ways that those who
hold them claim inspire them to engage in prosustainable bodily behaviors.

Indeed, the indigenous concepts of bodies, kinship, and reciprocal intersubjective relations with other-than-human persons (Morrison 2000) that
underlies the new animism points toward similar territory as that of the nonhuman turn of recent Western scholarship. This is because "The injunction
that emerges in zoöesis [i.e., bringing animal studies to drama and theatre to
move beyond the human in performance spaces] . . . is not that we refigure
the category of animal or understand the continuum between animal and
human; it is that we refigure the categorical thinking itself and, beyond this,
establish links to the biome that are not predicated on by-now obsolete definitions of human-animal-plant-mineral grown from the substrate of a Western,
positivist-inclined philosophical tradition" (Kubiak 2012, 55) that still largely
permeates the academy and thus religious studies.

The Australian environmental humanists Thom van Dooren and Deborah
Bird Rose attempt to bridge the gap between nonindigenous forms of new
animisms that move away from the above substrate, and bio-ecologies of
place, by framing the cultivation of an ethos of place through what they
call "ecological animism." This is a form of "relational awakeness" where
ecological animism "responds to a world in which all life . . . is involved in
diverse forms of adaptive, generative, responsiveness . . . life is saturated in

diverse forms of purposeful attentiveness and responsiveness" (2017, 258). Because human animals are part of life, we too are saturated with purposeful attentiveness and responsiveness. Our attention is noticing climate change—scientifically we are attuned to this, but many people in bio-ecological places are now feeling it with their own senses. This bodily experience, and the underlying concern for what it signals (a possibly maladaptive earth) is triggering responsiveness, including bodily religious dramaturgical responses. Chapters 6 and 7 will explore brief case studies of this emerging edge of religion and the body, and this ecological animist lens can apply to some of the case studies shared there.

Thinking through dramaturgical religious bodies in bio-ecological places can equally be aided by bringing to views of the body both diachronic (how something evolves over time) and synchronic (how something exists at a single point in time) understandings of evolution (Taves, 2010). From a diachronic perspective, as shared by religious studies scholar Ann Taves in her exploration of how the natural sciences can help scholars better understand religious experience/s, "At the genetic level, we share much in common with other animals; at the biological level, we have much in common with other mammals; and, at the social level, much in common with other primates" (Taves 2011, 173). Meanwhile, from a synchronic level, "human behaviors are simultaneously biological, social, and cultural. Moreover . . . within limits, the cultural and social aspects of human behavior can and do shape us at a biological level" (Taves 2011, 174). This calls us back to my earlier discussions of a soft naturalism at the meso-level, which privileges culture and social interactions, but a hard naturalism at micro and macro-levels. Yet Taves offers important insights for hard naturalism, positing that "in so far as cultural information is learned and neurons interconnect at molecular and cellular levels in ways that reflect what we have learned [via socialization and upbringing], cultural patterns are expressed at the micro level of neural processes as well as at macro levels of analysis, such as organisms and groups" (2011, 175). In short, it is impossible to disentangle human experience, and the entirety of our bodies, from the evolutionary processes that have shaped us as organisms, at both micro and macro levels, and vice versa. The various knowledge systems I draw upon for this book, from IK to QT to posthumanism to the nonhuman turn to performance studies to biocultural anthropology to religion and nature theory, all recognize our embeddedness as evolved beings; all are recent and needed forms of scholarship that critique the assumption that humans are the only material actors in culture and on the planet.

This methodological and analytical view has been richly put to use by the religious studies scholar Robert Fuller and his use of evolutionary sciences to rethink the religious history of North America (2013). As he explains,

"We now know that human beings are not blank slates. Our bodies bring distinctive biological and psychological characteristics *to* experience. Much of how we think and feel is biologically grounded and is not fundamentally affected by cultural discourse. The body is at the heart of all human thought and action" (2013, ix). Lest the reader think Fuller is a full reductionist, Fuller does explain that "Because our brains are not exclusively controlled by the information encoded in our genes, they require extrasomatic sources of information. Culture supplements our genetic codes in ways that make possible systems for gathering or processing information that far exceed their strictly biological foundations" (2013, 6). For my purposes I am interested in how culture, via religion, is responding to both the somatic and extrasomatic sources of information about changing planetary climate regimes, with this response in part being enacted through bodies via an "anthropology of everyday life [built upon] everyday sensibilities" (Seremetakis 1994, vii) of an observably changing biological world.

Given we are culturally-attuned biologically evolved beings, the impact of narrative and story on our understanding of observable (and nonobservable) reality is of import. Our evolved bodies belie a "basic morphology and physiology [that] limit the range of learned behavior. Animals, as products of natural selection, are also neurologically 'hardwired' in ways that restrict their learning" (Thiele 1999, 19). This hardwiring of bodies means that "genes determine the boundaries within which animals may behaviorally adapt to their environments" (Thiele 1999, 20). Given that all future adaptation of human bodies will occur on a planet above 415ppm of CO_2, and that at some point in time will most likely be devoid of permanent ice caps, an Amazon rainforest, and be home to rapidly melting permafrost, to name a few, how we behaviorally adapt to this home planet will become the pressing social, political, and economic question facing our entire species. It will also be a question addressed through religion, and because "it is cultural instruction, not natural selection, that serves as the primary engine of adaptation for modern humans" (Thiele 1999, 20), then adaptive cultural instruction via religions will be a determining factor in our species' ability to adapt or not adapt in ecologies of place to a climate-changed home. Can religious teachings and stories and ontologies encourage our bodies to create regenerative, resilient, adaptive lifeways in ecologies of place? If they can, and the evidence to date is that, at least at scales, they cannot (Snarey 1996; Taylor et al. 2016a; Wexler 2016), then the "moral dispositions" (Thiele 1999, 8) they can carry to and in our bodies will be "grounded in internalized stories. Narratives serve as the banisters of ethical life" (Thiele 1999, 8).

The stories we tell ourselves, and the ones in this book I am interested in are academic stories and religious stories, both, and how these impact our understandings of self, of community, of moral care, and of proper or improper behaviors, are central to our evolved capacity to live within planetary limits. Professional storyteller and scholar David Williams explains how storytelling is mental—our brains interpret sounds, place them in an understanding of language that has its own linguistic rules, and connect these with memories and an appropriate cultural context, and temporal—they take place in places and reference places. While listening, "our brains are continually reassessing the entire narrative paste" (2012, 99), and emotions and psychological dispositions are triggered. If our stories are devoid of place, then our ability to respond to the changes of place is severely limited. We do not see how we are "stalked by stories" (Basso 1996). If our stories only focus on human actors, then we are not prepared to understand the agency of nonhumans and how we must respond to this agency. If our stories abstract out the human—whether these stories are economic, political, ethical, legal, academic, or religious—then they do not prepare us for being properly placed within a larger planetary home that is alive and that acts upon us, in ways that are nonlinear, uncertain, discontinuous, and unstable (Gunderson and Allen 2010, xiv–xv). We need stories, and thus scholarly studies, that recognize Western ways of knowing nature have been in many ways maladaptive (to be clear: ways of knowing nature from other cultural stories have also been maladaptive [Tuan 1970]), are not the only way of knowing nature, and that recognize we are always engaged in "intimate encounters" (Bender 2002, 136) with agental others that engage all of our senses. We need stories and studies that account for the evolved aliveness of our bodies, and thus of culture, and how we are always already nature-bound beings, now having to navigate nonlinear cascading effects of rapid climate change. How do our religious bodies, and the stories they shape and are shaped by, navigate this reality? The case studies presented later in this book provide glimpses-in-time of some possibilities.

CULTURAL LIFEWORLDS

The anthropologist Michael Jackson states that "any social order is permanently under pressure to accommodate, contain, or curb potentially divisive needs" (2017, 15). This is even harder in an increasingly globalized world that privileges individual choice and what I call the tyranny of mobility—accessing dense energy in the form of fossil fuels to easily leave homeplaces and communities and forge new social networks and identities, often devoid of connection to what I call ecologies of alive places. Historically religion has been a carrier of social order, but increasingly religious identities as a subset of culture contribute to divisive needs. This is being witnessed in real time in the United States over issues related to gay rights, abortion, gun control, and responses to climate change. These divisive needs will most likely become more pronounced as humans struggle to square away some cultural stories and myths (e.g., the economic axiom that nature has no value until human labor is added to it, or that we are individual profit maximizers) with emerging ones (we need to reduce CO2 emissions).

This process of navigating changing ecosystems will impact cultural narratives, rites of passage, language and communication, understandings and use of technologies, kinship identities, and mobility/migration. It will also impact actions and experiences unique to religion: pilgrimage, rites of passage, institutional teachings and doctrines, and material practices and products. The pathways these cultural and religious changes can take are constrained in part by the cognitive scaffolding of our minds, but also by "conscious reasoning and reflection [that shapes] cultural innovation and transmission" (Whitehouse 2007, 250). Teachings from elders, community leaders, parents, education networks, and the media all play a role in the latter. However, these roles are in flux in a world of competing messages about climate change; or with elders whose lived memories and experiences of place do not reflect the weather patterns of today, impeding their abilities to help provide adaptive insights to communities-in-places. Certain rituals, especially religious ones around agriculture or rites of passage that are seasonally specific, may not carry forward and have the same explanatory power or cultural significance that they once carried. What does it mean to worship a deity with the first spring thaws, for one hypothetical example, when there is no longer a deep freeze such worship traditionally follows? What then of traditional cultural and thus bodily obligations? How might bodily religious dramaturgy respond to agental earth shifts? What cultural teachings and mores will be encoded in such responses? How may these shape our emplaced actions in biocultural contexts, especially if they inspire or support prosustainable behaviors in response to rapid climate change? These are important questions facing our field, and our scholarship must attend to them. I turn now to the second half of the book where I shift from building an analytical lens and argument and move to explore answers to some of the above questions through analyzing various case studies. I begin the second half by first turning my attention to cultural narratives of "oil culture" via new materialist and energy humanities lenses and how these narratives, and the politics they buttress, must be factored into religious studies as well. This is because the performance of religious bodily dramaturgy for most humans the world over is currently afforded by the agency of ancient forms of sunlight, the exploitation and use of that also drives global warming.

NOTES

1. But even when understanding "nature" in this sense we must be leery of scientific fundamentalism, which too often shuts down understandings of agental forces and agental entanglements.

2. On this, see Joan Roughgarden's 2009 *Evolution's Rainbow: Diversity, Gender, and Sexuality in Nature and People.*

Part II

APPLIED CASE STUDIES

Chapter 4

Liquid Black Death

A Hegemon Ancient and Seductive

The purpose of this chapter is to help understand how material effects and agencies of fossilized forms of ancient sunlight have come to structure the economics, politics, and cultures of most humans the world over. Relatedly, its goal is to help further reflect on how the use of this ancient sunlight is overdetermining biogeochemical systems and bio-ecologies on our home planet. In short, to understand the dramaturgical bodily enactments of religion in an era of climate changes we must understand the to-date key driver[1] of these changes, which is the use of liquid petroleum and its distillated products[2] at mass scales. If we are to have any hope of stopping runaway climate change we need to understand the key driver of such change, and thus better study those within religions who are responding via bodily dramaturgies to the same change. More and more religious production and bodily practice by the 2030s and 2040s will likely be in service of stopping, or at least responding to, such climate change and thus the causes driving such change. And as I will return to in the book's afterword, academics need to understand what is driving climate change (and its connected siblings species extinction, deforestation, loss of topsoil, loss of freshwater) so we can both see how as academics we are complicit, and how we need to collectively join in using the academy to proactively stop what is driving climate change.

A key insight about seeing the use of fossil fuels as a key climate change driver is that "one of the uncanniest effects of climate change [is a] renewed awareness of the elements of agency and consciousness that humans share with many other beings, and even perhaps the planet itself" (Ghosh 2016, 84). In bringing religious studies and environmental humanities to understand the significance of climate change, we must bridge the gap between assuming liquid petroleum to be inert matter that is background and instead shift to understand oil as agent of performativity. We must move beyond the gaze of the humanistic Abstracted Ego and Isolated Subject of Mastery-Over an Inert Planet, and instead with our scholarship enter into theorizing and analyzing the many "ontological entanglements" and "ontologies of immanence" (Asberg et al. 2015, 162, 145) that shape our lifeways. We must see "a multiplicity of prominences [and] work for alternatives to dualistic splits in epistemology" (Asberg et al. 2015, 150). And in this, to understand performed religious lifeways, we must understand as well oil lifeways, and how oil is an actor in such lifeways. As the Petrocultures Research Group in Canada explains, those in the developed world live within petroculture: "the ways in which post-industrial society today is an oil society through and through. It is shaped by oil in physical and material ways . . . fossil fuels have also shaped our values, practices, habits, beliefs, and feelings" (2016, 9). Religious bodies in metropoles are thoroughly shaped by petrocultures, while those at poles are increasingly shaped by petrocultures, if nothing else because biocultural lifeways at the poles of empire are being impacted by the climate changes triggered by those living in metropole petrocultures. This also suggests that resilient, adaptive (new) religious dramaturgies will most likely have to decouple from petroculture, at both poles and metropoles.

To begin, though, let us journey briefly back to the year 1859. This year was pivotal for two unrelated reasons. The first was Charles Darwin's publication of *On the Origin of Species*. This became the first published and widely read factual basis for a naturalistic understanding of the natural world and all observable phenomena. This book you are reading is in the same intellectual heritage, utilizing naturalistic understandings of our biophysical bodies as embedded in bio-ecosystems while also theoretically adopting a chastened materialistic understanding of religious phenomena. The latter recognizes religion is produced by social groups consisting of bodies-in-dramaturgical-performative-action, with that action in agental relation with more-than-human others in bio-ecological places of dwelling.

The other key event of that year was the opening for commercial purposes of Drake's oil well in Pennsylvania. By tapping into liquid black gold this business began the shift in the Industrial Revolution from coal (still used to this day by humans the world over for heat and to generate electricity) to liquid petroleum, or what we commonly call oil. Fast forward 150 years and the entire world is saturated in the residues of using such liquid form of ancient sunlight: gyres of plastic products throughout the oceans; CO_2 trapping incoming sunlight energy as heat in our atmosphere; oil byproducts in the mother's milk of mammals; and when scraping the bottom of the barrel of refined oil, the asphalt of roads that now extend deeper and deeper into the last remaining biodiverse terrestrial hotspots on the planet. Even the computer at which I type this book, and the glasses on my face to see the screen, to the vegan tofurkey I ate for Thanksgiving this year—it's all a product of oil: from extracting, to manufacturing, to shipping, to consumption, to throwing it "away." It's all powered by oil, in an assumed linear economy built on models and thus policies that externalize all racialized and gendered social and environmental trauma that such oil dependency engenders (Guha 2006; Roberts 2007; Mander and Goldsmith 1996; Turner and Clifton 2009).

We can also add natural gas to this category, as it functions the same as oil in terms of providing energy to the industrial grid and comes largely from similar fossilized sources of old sunlight. Here remember that refined oil and gas products are the fossilized bodies of plants and ocean-based organisms that photosynthesized sunlight hundreds of millions of years ago. Natural gas is turned into energy (think the natural gas received from the process of hydrologic fracturing of rocks to access natural gas reserves, or "fracking"), but more importantly is used as artificial fertilizer via the Haber Bosch process to provide artificial nutrients to crops. This process has allowed for the Green Revolution to create excess calories for large numbers of the overall 7.5 billion human animals extant on earth. Here too the models of a linear economy[3]—fallacies of misplaced concreteness at their worst—bely the true cost of using such natural gas products on the human and nonhuman bodies found in agricultural systems the world over. This also leaves out the political economy questions of food waste, food apartheid, land clearances, and other maldistribution structures of Green Revolution lifeways.

The agricultural geneticist Wes Jackson explains how the above layering of biosocial lifeways on old liquid petroleum are a result of the "Five Pools of Carbon" (2010, especially chapter 4) humans have tapped into and that have enabled us to live beyond the sustainable bioregenerative capacities of the earth. Jackson explains how this occurred when our ancestors transitioned from gathering and hunting for their sources of calories—living off of contemporary sunlight—to settled domesticated agriculture, when about 10,000 years ago we began to live off of ancient sunlight stored in soil carbon. This Agricultural Revolution allowed the human animal to tap into ancient forms of carbon via mining it from the soil (the first pool of ancient carbon) and began the march of our species away from true sustainability. We are in the throes of this Revolution to this day.

The second pool of ancient carbon was accessed via old-growth forests when these forests were felled to fuel the Bronze and Iron Ages. The clear felling of these forests powered the expansion of Empires from about 5,000 years ago through the Middle Ages. The third pool of carbon layered on top of these two, and that is when humans, largely in Europe and Asia, tapped into old forms of sunlight stored in coal. Here in Western Europe the Industrial Revolution began, where access to coal coupled with technological innovations helped enable rapid industrial production of goods and an oceanic going mercantilism (Vries 2001, see esp. 436–440). In this Revolution the use of coal was coupled with violent land clearances, the rise of industrial cities, the (ongoing) trauma of White Cishetero Settler Colonialism as European powers violently expanded throughout the world, and the development of what is today's neoliberal capitalism. The natural world becomes irrevocably changed, as well, as these old pools of carbon are tapped into and our lifeways become structured around them.

The stage is set, however, for the final two pools of carbon. The use of these pools are coupled with penicillin and other modern health care inventions that combine to allow for exponential human population growth, to go along with exponential growth in human power via technologies to drastically shape the earth's biomes. These final two pools Jackson discusses are accessed by tapping into the stored carbon of oil (pool four) and natural gas (pool five). This also allows for the Green Revolution of natural gas-based agriculture, and for the massive altering of the chemical construction of our atmospheric commons. The agricultural products derived from natural gas and hybrid seed technologies also allow for concentrated animal feedlots and a massive boost of methane in the atmosphere. The use from Drake's first oil well[4] to the use of petroleum from contemporary oil fields of the North Sea, Gulf of Mexico, Russian steppes, Middle East,[5] Texas, Venezuela, Nigeria, and other key petroleum producing areas of the globe has marched us into the Anthropocene. This history, presented here as a linear march of time, is in many ways precisely such a linear progression. Indeed, professor of International and Public Affairs with a focus on the politics of energy Timothy Mitchell reminds us how both oil-consuming states and oil-producing states are now, 150 years removed from Drake's prospecting, wedded to a "carbon democracy" (2011) that thoroughly shapes the economies, supply chains, and political constraints of building our globalized civilization around oil.

And it is easy to see why humans so readily utilize liquid petroleum, despite its impacts on bio-ecosystems: one gallon of oil translates into approximately 500 hours of human labor. Why would we not want to have an easier life, with more leisure time, powered by dense energy, each with our own individual 2,000+ pounds of metal to drive around in to pick from one of ten different types of laundry detergents at the nearby market? However, as Wes Jackson also points out, we are going to have to be the first species to willingly turn away from old, dense energy and say "No." That is the challenge of our time, despite that "[f]or the past 3.45 billion years nearly all life forms that we have observed have depended on energy-rich carbon as a fuel source" (Jackson 2010, 73). Despite, too, that the products of liquid petroleum and its distillates shape every aspect of our modern lives, from kerosene to jet fuel to natural gas to plastic products to fiber for clothing. This total cocoon of oil around our contemporary lives gives a different truth to Bruno Latour's argument that we have never been modern—we have never been modern because we are living off premodern energy. Yet, it's precisely the burning of this old dense energy that is altering the planet into a stochastic planetary regime shift: one that will be maladaptive for our survival, and which will overdetermine the bodily dramaturgical production and practice of religion in bio-ecological places in the years to come.

OIL AS LIFEWAY

There's another layer, too, to petroculture: a performative rupture in the composist terraworlding of bodily becoming. This layer is that "in looking for sources of climate change in human subjectivity, we need to take distracting excess—of fossil fuels, greed, whatever—for granted. Might humans then realize instead the *subtly* refractory, recalcitrant and unexpected toxicity of our simplest normal daily performances?" (Kershaw 2012, 7, italics in original). This subtle recalcitrant and unexpected toxicity upon the symbiotic real in ecologies of place is a hidden cost of scaffolding religious bodily dramaturgy on the coal, oil, and natural gas pools of carbon. Yet this cost is no longer hidden, as all bodies everywhere are experiencing bio-ecological shifts in places due to 200 years of distracting excess. How will religious bodies respond? This sighting of possible and likely futures must also account for post-Anthropocene settlements that have much less mobility, therefore resulting in a radical alteration in size and layout of settlement patterns and their locations—and thus how religious practices will be bodily performed in post-Anthropocene settlements.

Given the postanthropocentric lens of this book we must also understand how constrained post–Peak Oil and post-Anthropocene flows of liquid oil and human bodies in an era of climate chaos will impact academia. As academics we need to understand that our entire professional project is scaffolded upon and enabled by use of dense forms of old sunlight. This literally means that academia as we know and practice it is by bio-ecological and biogeochemical definition, unsustainable. It also means that the subjectivities and logics we perform and enact as professionals are made possible in large part by oil and coal. Oil and coal are the material agents that power the bodies of our students as they transit to campus; that provide the fuel that gets food to our campuses; that power the energy grids upon which our buildings and servers depend; that transport us to conferences and research sites; and that power the reward system of tenure, where with more money as we advance through rank[6] we can enter deeper into the consumer economy. Yet as the political scientist Joseph Mohorčich explains,

It is impossible to know the specific effects, challenges, and constraints that will accompany any exit from the human-dominated geological epoch. Most plausible post-Anthropocene scenarios, however, envision a decline (slowly or not) in human control of the material and surface area, and therefore the energy stocks and flows, of Earth. This potential reduction in the total amount of energy available to human societies will constrain the ability of these societies to move people through physical space, a task whole scale and complexity in existing societies remains significant . . . transportation after the loss of current levels of energy availability is likely to be slower than existing transportation, even holding mode choice constant, because of the cubic relationship between velocity and the power needed to overcome drag. (Mohorčich 2020, 2375)

The implications of all of this for the academy (travel for research, for conferences, for sabbaticals, of geographical reach to recruit students) are profound and currently not being discussed, at least in any serious way about what education in a post-Anthropocene setting will resemble. This is to our collective detriment and signals a severe lack of responsibility and foresight we are abdicating as privileged educators capable of futures thinking.

Within this context I must also by default explore issues of power, for once we look at fossil fuels as driver of climate change and the politics of, we are in a geopolitical context of resurgent Settler Colonial patriarchy and misogyny. As the historian Timothy LeCain writes, "when humans use things like coal or oil to generate social power, these things demand that humans conform to their material needs, thus shaping the way power is created and exercised" (2015, 17). This is especially true for how indigenous peoples are treated where such peoples live in areas that have fossil fuels under the ground or under bodies of water, and such an investigation into the impacts of fossil fuel exploration on the religious practices of indigenous communities is a subject worth of further study.

Here, though, I return to a key insight of material feminism: other bodies have agency. These evolutionarily evolved bodies include flora and fauna (and the fossil fuels they have become), as well as microorganisms and fungi and bacteria, but also ecosystem-scale geological agents like rivers, glaciers, rainforests, estuaries, and seasonal changes. In short, the human animal is shaped by active more-than-human agents, and we shape these agents, at individual to aggregate ecosystemic scales. We are a bio-ecological, biocultural sympoeitic world braiding of composist terraworlding. And with climate change, this sympoeitic mutual intersubjective becoming is now thoroughly global.

As explained by Karen Barad, material feminism shifts our theorizing from a static, boundary defined, representationalist perspective that assumes an ontology of "independently existing entities," to one of posthumanist performativity. As she argues, a performativity lens advocates for a "materialist, naturalist, and posthumanist elaboration that allows matter its due as an active participant in the world's becoming, in its ongoing 'intra-activity'" (Barad 2008, 122). Such a perspective is valuable for examining the energy flows of the earth, including the form of energy flow that has most overdetermined the political, economic, and ecological reality of the Anthropocene epoch in which we find ourselves: oil. It is the mattering of the liquefied bodies of old plants and sea organisms that when combusted are changing the climate of our time. These ancient bodies through their burning are active participants in the world's becoming, where through the intra-active agency of their chemical molecules they trap the sun's heat. And it is the trillion-dollar military-economic quest for such decayed and liquefied bodies that shapes the traumatic dispossessive settler politics of our time. And it is the dramaturgical practice of bodily religious production within the above realities that are the religious practices of our time.

In short, and I am assuming this is the case for readers of this book, it is as impossible to imagine our lives without oil in them as it is to fully comprehend how insidiously wrapped up with oil our lifeways truly are. Petrotechno imaginaries have captured our attention and sense of self, even if they have become backgrounded or strategically hidden through repeated encounter (think endless television commercials for cars), planned routes of extraction and manufacture (Szeman and Whiteman 2012), or the inability to see oil "as giving shape to the social life that it fuels" (Szeman 2013, 146).

Meanwhile, the actual use of oil in our individual post–Industrial Revolution lives has been rendered invisible such that the task is "to understand, given the centrality of oil, its weird natural/cultural status, what oil has been in history, and what it will be: politically, culturally, aesthetically, historically" (Stoekl 2014, xiv). Given the overall centrality of oil to everything humans do, and at the very least the impact on the biogeochemical cycles of our planet, then to study religion adequately in this epoch we need to see that it is "problematic *not* to include energy in our narratives of historical change and development, including social and cultural shifts and transitions" (Wilson et al. 2017, 5).[7] Thus, to understand the social and cultural shifts of religious dramaturgical bodily practice in an era of climate changes, we must understand energy sources and supplies; and as long as it is still the dominant form of unsustainable energy use, we need to be aware of how the biocultural behaviors we study are largely made possible, often through hidden ways, by the material agency of combusting oil. In some of the case studies to follow I make such connections to oil specific; in others, the oil is hidden, but it's material agency is still one of the factors driving the dramaturgical responses to climate change (and other bio-ecological harms) I explore. Regardless of whether oil is foregrounded or invisibly giving shape to the social lives we study, we need to understand the looming "disaster of the end of oil" (Szeman 2007, 815). This disaster is already bio-ecological, and will unfold to be economic, as well, impacting religious bodies, and the bodily practice of academia, both.

PERFORMING THROUGH OIL

In this section I want to tease out my first case-study analysis, one that explicitly explores sympoeitic oil terraworldings. The case study fuses religion and protest within the techno-imaginary of oil democracy and empire. It is offered here as a way to begin thinking through specific instances of religious bodily performance in an era of climate changes that subsequent chapters will undertake. The immediate exploration is inspired by a call for papers from the Feminist Theory and Religious Reflection Unit for the 2018 American Academy of Religion Conference, to which I submitted and presented on what follows. The call specifically asked papers to reflect on the trope of "energy," and how this can be "a potential metaphor for understanding, advancing, and identifying the various ways religious feminist work transforms and energizes the public sphere."[8]

I begin by combining Barad's insight from earlier in the chapter—that matter is performative—and bring this to the "dark matter" of oil as articulated in the 2017 documentary *Awake: A Dream from Standing Rock*.[9] In this documentary filmmakers bring viewers to the #NODAPL protest camps at the Standing Rock Sioux reservation (North Dakota and South Dakota, with the base camp in North Dakota) where in 2015, 2016, and 2017 water protectors labored to defend the sacred waters of the Missouri River from the "black snake" of the Dakota Access Pipeline. For water protectors this snake provides one foil of "dark energy" (vs. the clean, sacred waters of the river) and represents an addiction to fossil fuels, empire, and the violent subjugation of the earth. Protestors, led in large part by Sioux and other indigenous elders at the Standing Rock base camp, used prayer, fasting, chanting, and other forms of ceremony to both symbolically and physically stop the installation by Energy Transfer Partners of a pipeline across lands and waters held sacred by Plains Native Americans, especially by the Standing Rock Sioux.

Statements of such sacrality are found, for example, on the website Stand with Standing Rock, created and maintained by Standing Rock Sioux members, which states "In honor of our future generations, we fight this pipeline to protect our water, our sacred places, and all living beings."[10] These views helped inform dramaturgical ritual actions at the off-grid, collectively organized and managed base camp: from sage smudging as people gathered in ritual circle; to prayers and offerings made to the directions, to ancestors, and to water protectors at the base camp; to host drum circles with sacred songs and community dances late into the night; to offerings of tobacco and other sacred plants; to prayers made in and to the river; to prayers and offerings made in efforts to convince the militarized security presence to leave sacred territory—these performative actions created the context of religious practice at Standing Rock. As indigenous youth activist Danny Grassrope of the International Indigenous Youth Council explained, "We've got to stay peaceful and stay in prayer because this is a ceremony and we have to protect what we have here, and that's our prayer."[11]

In looking at the demographics of those at Standing Rock[12] it should be appreciated that for many indigenous views, broadly speaking, the understanding of nonhuman bodies and powers influencing human religious bodies and ceremonial life is long-lasting, and includes "pre-patriarchal and pre-colonialist indigenous [understandings of] nationhoods" (Guerrero 2000, 44). These nationhoods are peopled by other-than-human persons with their own cultures, so that a key purpose of ritual, ceremony, fasting, dance, and other technologies of the indigenous religious body is to help "maintain . . . relatedness with neighboring others" (Bird-David 2000, 96). Such insights informed the dramaturgical practices of those at Standing Rock, as the elders and indigenous peoples present remained in cultural pathways that never stopped practicing and knowing the themes that Barad presents above. And key is that those on such a respectful relational biocultural pathway successfully held off "big oil" through ceremony, nonviolent protest, and active community building at the base camp. #NODAPL provides evidence of a mix of established practices, as this respectful relationality is something that was never lost from Sioux peoples, broadly speaking—theirs is a world populated by vibrant, living, agental forces and powers present in the visible bio-ecology within which all beings dwell—with new religious production: that these forces and powers are sacred to the Sioux was the guiding inspiration for protecting their sacred waters, burial grounds, and landscapes, where this protection became the ground (literal and figurative) of contemporary religious production driven by an awareness of the harms of petroculture. The generationally held and lived understanding of a posthuman relational worlding was therefore present at Standing Rock, informing religious dramaturgical production in response to the violent logics and real-world impacts of continued fossil fuel use.

Significantly, at least as presented in the documentary, the response to this violence was one of nonviolence. The nonviolent response, tightly tethered to religious performativity, was led by female native elders and also native youth, both male and female and non-binary, who were there to protest both the desecration of a sacred River and the lands around it, but to also bring about the healing of their own culture. Standing Rock was an alcohol-free, nonviolent settlement, where those present prayed to and conducted ceremony to a living, dynamic, energetic sacred earth each day. Within this context the black gold that was being extracted and transferred by the pipeline was equally of sacred worth to a dynamic earth, but in the sense that it should remain underground, part of the natural ecosystem where it is perceived to naturally belong.

From Barad's perspective, it is evident that the nonviolent ritualized prayers and protests of those at Standing Rock, undertaken in defense of for those present the sacred flows of water, the sacred flows of seasonal change threatened by climate change, and of the sacred flows of the landscape upon which the Standing Rock Sioux reside, are a form of discourse. As she says, "Discursive practices define what counts as meaningful statements" (Barad 2008, 137). And in this context, pace Barad, "'Humans' are neither pure cause nor pure effect, but part of the world in its open-ended becoming" (Barad 2008, 139). The humans at Standing Rock as shared in this documentary were in open-ended becoming with the hill next to the River Missouri that they climbed for prayer; were in open-ended becoming with the frozen water formed on the bodies of native youth who were sprayed by police and private security forces during a stand-off at the bridge on a freezing night, where such intersubjective agental becoming resulted in the flu for some sprayed; in agental becoming with petroleum-become-rubber shot by the agent of gunpowder and human finger and melted minerals in the form of a gun the open-ended performative blunt force trauma such form of rubber had on the face of a female reporter; and in performative dialogue with the River itself that those at Standing Rock bathed in, swam in, and paddled across with canoes and kayaks. The humans at Standing Rock were also in open-ended becoming with the black liquid that coursed under the ground, itself an agent that shapes the ongoing politics, economics, and military interventions of carbon democracy, and that demanded of those at Standing Rock a resistance to the logic of domination (see below) that says such liquid gold is inert, agentless, and there for instrumental human use that goes to benefit, at least economically, the world's elite. So all this is a dialogue of materialist feminist flows where the energy flow in this case study is that of liquid—both of a River, and the liquid of fossilized bodies of carbon found under the soil, and of the female bodies of human animals united in ceremony and prayer to defend such sacred energetic flows.

From an ecofeminist lens the violent dispossession inherent to the mining of and prospecting for the black gold of fracked natural gas, shale oils, and other forms of fossil fuels is a carrier of violence against these same women's bodies and the earth. The violent governmental response, at least at the level of the States of South and North Dakota and also by regional law enforcement and private security firms employed by Energy Transfer Partners, evidences a "logic of domination" (Warren 1990), where white, male, human, and capital are bounded domains of identification that are superior to their respective binaries. In this case those are bodies of color (especially indigenous bodies), female, nonhuman, and noncapitalist on the lesser side of their respective binaries. The logic of domination operates such that those in power are justified in violently subjugating the latter bodies and forms of agency, with this subjugation sanctioned by patriarchal state proxy. An ecowomanist lens (Harris 2017) would add to this analysis that it is the violent subjugation of bodies of color—here, those of indigenous women in the ongoing process of genocide that dates back 500 years—that exemplifies the prospecting for liquid energy flows as standard operating procedure for White Settler patriarchal colonialism. That water is life—the most essential material form of flow for terrestrial becoming—is weaponized and threatened, and denied to indigenous female bodies of color—is further evidence of the violent subjugation of the agency of such female bodies. When theorizing dramaturgical religious production on a planet going through climate shifts, and in an academy wedded to petroculture, we have to make sure our analyses attend to these forms of violence so we can work in solidarity to stop them.

OIL, RELIGION, AND VIOLENCE

Along with historic cases such as #NODAPL, it is important to consider possible future scenarios of oil depletion coupled with runaway climate change. Such scenarios need to be taken seriously by the academy. And in this I want to focus on one behavioral trait our species too often exhibits, which was sadly but predictably all too present at Standing Rock by those acting on behalf of petroculture: violent confrontation. To understand religious dramaturgical bodily production and performance in this era of climate changes we need to spend some time with a sober assessment of how humans will likely treat one another and who will be the haves and have nots in the years and decades to come in various ecologies of place. We also need to understand how who the "haves" and "have nots" are will impact bodily religious practices in the Anthropocene.

The performative leverage point of oil culture on biogeochemical regime shift means that "a global landscape of Anthropocene inequality is coming into being" (Purdy 2015, 6). This inequality is being driven by land grabs, biopiracy, continued genocide of indigenous peoples, structural racism, sexism, and queerphobia, and the military interventions required to keep carbon consuming economies/countries at the apex of the carbon food chain. As human actions drive average global temperatures well past the maximum 1.5 C rise above pre-industrial baseline agreed to at the Paris climate conference in 2015, we should expect to see major socioeconomic-ecological triage over the rest of this century. Water shortages, droughts, decreased food production, viruses on new vector paths, stagnating economies, massive ecorefugee migration—all of these are predicted impacts of runaway climate change. The studies and news reports that adumbrate these and other Anthropocene futures are so numerous that it is pointless to even cite a few here to help further make my point. We all know what the models predict and should know what this means in how we actually get our calories, get potable water, move our bodies from point A to point B, and how we engage our body politics. As Karl Bonnedahl, lecturer of business, economics, and statistics, and Pasi Heikkurinen, lecturer in business and sustainable change argue, we need to rethink sustainability and thus our economics, ethics, and lifeway practices on a planet that is only going to get hotter and fuller (Bonnedahl and Heikkurinen 2019). And as per one of the arguments motivating this book, we need to rethink our educational and research practices, and how we'll embody them in changing bio-ecological habitats.

The science of complexity, and the insights of relationship, connectivity, interdependence, self-organization, and stochastic nonlinearity germane to this science (Capra and Luisi 2014), is a helpful lens for understanding the end of oil culture on a hotter, fuller planet. Culture here is the key point— culture, as defined earlier, as something created by organisms living in community, is responsive to larger biogeochemical shifts. Because human biocultural evolution describes culture as coupled with biogeochemical systems in bio-ecological places, we must recognize culture can be and is a driver of the shifts in the latter. However, the material impacts on these systems by culture-bound organisms, of which human animals are one type, is only one aspect of the biogeochemical package of the planet. Because culture is a co-creation of evolving organisms in evolving bio-ecological places, evolved human biocultural lifeways are always dependent upon the larger biogeochemical systems and regime shifts in bio-ecological places of those same ecologies of place within which they reside and interact. These regime shifts "can be traced to a small number of variables" (Gunderson and Allen 2010, xvii), which in the Anthropocene are the heat-trapping properties (material agency and effects) of greenhouse gases, nitrogen deposition in the soil, and ocean acidification. These are interlinked variables, driven largely by the human use of oil, and while other variables will and do impact the regime shift of global climate change, those listed directly above are key oil-based drivers. Almost every model to date is clear that the impact of these variables on regime shifts will cause loss of food, loss of freshwater, decreased levels of oceanic oxygen, shifting ocean currents, retreating ice caps, decreased soil fertility, and other system responses equally inimical to the healthy flourishing of human biocultures. It is for these reasons that LeCain posits that "we may have to undergo a very difficult process of disentangling ourselves from some very powerful material things that have increasingly come to dictate our collective fate . . . humans and their cultures are best understood not as the creators of their destiny and environment, but as products of a material world that is constantly creating and recreating them" (2015, 22–23). It is also one of the reasons he argues that the term Carbocene (also Carboncene) is more appropriate for understanding the unfolding geopolitical epoch, as compared to the Anthropocene, in part because it recognizes "the powerful co-starring role played by coal and hydrocarbons like oil and gas in creating our current era" (2015, 23).

A recent publication hints at the severity of what we as biocultural animals-with-religion who are products of a material world face:

> Regime shifts occur across a wide range of social-ecological systems. They are difficult to predict and reverse and often produce sustained shifts in the availability of ecosystems services. When a system undergoes a regime shift, it moves from one set of self-reinforcing processes and structures to another. . . . As humans increase their pressure on the planet, regime shifts are likely to occur more often and more severely. (Rocha et al. 2018, 1379)

This insight is shared within the authors' larger comparative study of 30 biocultural regime shifts across various biomes that tested for domino effects and hidden feedbacks. In it the authors caution that ecosystem changes around the world will most likely impact other ecosystems, and such nonlinear impacts will impact climate, moisture content in the atmosphere, agriculture, and flows of water, often in ways that will impact human wellbeing.

The implication of these regime shifts is violent conflict: "the wars of the future will largely be fought over the possession and control of vital economic goods—especially resources needed for the functioning of modern industrial societies . . . resource wars will become, in the years ahead, the most distinctive feature of the global security environment" (Klare 2001, 213). If societies can maintain any resemblance of cohesion at scale, especially during resource wars and threats of collapse, is an open question, with profound implications for the academy. This is an assumption that may not hold, though. It is very likely that societies organized around statism will collapse in the coming decades; or that fascist regimes will emerge based on ethnonationalist fervor, with some of these regimes possibly replacing the state, or acting de facto as the state. Choose the collapse scenario—whatever it is, it will likely bring both small and large-scale violence.

So while, for example, "only a high level of soil degradation increases the risk of armed conflict [such that there is] weak support for an environment-conflict linkage when tested with large-n methods" nonetheless our "focus [should be on] lower-level conflicts, in which the context of poverty [as the] social distribution [of renewable resources] is the most important factor in linking conflict to scarce resources" (Theisen 2008, 802). Here the cause of violence will be, yes, triggered at times by massive bio-ecological degradation, but more likely, at least in the present and coming years, localized violence triggered by poverty and scarce access to needed resources. What this conflict resembles will be regionally specific, due to factors of governance, supply chains, cultural practices, and place-specific bio-ecosystem flows related to the availability of renewable resources in a region and how many citizens of such a region can access them. However, this suggests that the reality of ongoing resource wars of different types and scales will become part of twenty-first century living. The availability of taken-for-granted renewable resources will likely become more stressed and social disparities more pronounced given regime shifts caused by our use of oil products.

When we add to the above insights those from Professor of Geography David Harvey, namely that "Spatial and temporal practices [are] closely implicated in processes of reproduction and transformation of social relations" (Harvey 1990, 2018), further cause for restrained caution is warranted. This is because we must begin to think about the coming changes in (religious) social relations as being driven in part by shifts in spatial and temporal practices due to the Anthropocene. Harvey writes elsewhere that "The history of capitalism is, then, punctuated by intense phases of spatial reorganization" (1996, 296). How much more intense may these self-organization phases be when we enter into post-carbon capitalism triggered by various resource shortages? The implications of such spatial reorganization triggered by changing Anthropocene economics, and for sure by changing biogeochemical cycles, are of a major re-worlding of human biocultural lifeways in bio-ecological places. This will impact the academy. And it also will include, as this book investigates, the reorganization of religious dramaturgical practices that occur through the enmeshed, entwined, performative bodies of human animals.[13] That such major spatial reorganization and displacement (Renner 2013) will be in large part driven by regime shifts both economic and bio-ecological suggests a context ripe for violence and conflict at various scales in a post-oil, resource stressed Anthropocene world.

This context will both inform and be shaped by religious actors. Here the caution of religion scholar Karen Armstrong is apropos, but also her larger point, as shared in her exploration of the history of religious violence in Axial age religions: "Even those who admit that religion has not been responsible for all the violence and warfare of the human race still take its essential belligerence for granted" (2015, 4). And just as Michael Klare soberly suggests resource conflict will be the norm as we collectively experience Anthropocene and carbon capitalist regime shifts, so too we will see this post-Agricultural Revolution history of post-Axial age religious actors as agents of violence continue. The potential for increased violence in an era of resource shortage will be exacerbated because "The problem of religious violence . . . has greatly increased in the current era of globalization" (Selengut 2017, vii). Add to this that the ability to cogently discuss real (vs. perceived) threats and strategize effective political solutions to climate change is currently lacking in the United States, in large part because of religious identity (Haidt 2012), and it is a tough predicament to be in. At the time when humans need to come together to navigate climate triage, and in so doing overcome the worst inclinations seen over and over again in the anthropological record of scapegoating, targeting minoritized peoples, and active physical violence during times of environmental stress, these (religious) divisions are not helpful. It is an open question that demands our attention, though, as to how humans will relocalize their biocultures in a future with much less fossil fuel use. Those of us in the global elite will not have the luxury of dense energy (made possible by externalizations and military might) provided by the material agency of old life forms. Economics, politics, education, religion, agriculture, health care, concepts of national security, statism, where people reside, and biocultural identities and practices from language to ritual to definitions of the family—all are going to change and likely become regional. This relocalization will most likely be messy, disruptive, and violent, but hopefully it will also at times and in places be adaptive and equitable.

I conclude this chapter with insights from three scholars that are useful for reminding us of the context of oil within which religious lifeways currently exist, and the context of the climate-changed Anthropocene in which religious lifeways will be performed in the years to come. The first is from the geographer Matthew Huber, who argues "that violence is *constitutive* of the crisis-prone and unstable capitalist market" (italics in original) (Huber 2011, 817). The second is from Professor of Development Studies Swapna Banerjee-Guha who explains that, "Globalisation [sic] can be characterised [sic] as a producer of extremely divergent trajectories of development and underdevelopment" (Banerjee-Guha 2010, 10). The last is from political theorist William Connolly, who opines that "Militant citizen alliances across regions are needed to challenge the priorities of investment capital, state hegemony, local cronyisms, international organizations, and frontier mentalities" (Connolly 2017, 8). We see that hegemonic oil lifeways are dependent upon and promote disparate forms of violence, and that a political response of creaturely solidarity in ecologies of place is needed to navigate these. How religious practice can contribute to this will be a focus in chapters 6 and 7. For now I turn to case-study explorations of bodily religious production in specific geographic regions already dealing with maladaptive regimes of biogeochemical and bio-ecological shifts triggered by the anthropogenic climate changes explored in this chapter.

NOTES

1. It is entirely possible, although highly unlikely, that many humans will completely transition to renewable forms of energy—hydroelectric, geothermal, tidal, solar, wind—although to scale this out to replace fossil fuels is currently beyond any capacity of micro- or centralized grids. It is also possible, and even more unlikely, that humans could transition to nuclear power (leaving aside the issue of dealing with toxic radiation). While the former is more realistic, and signs point that there is indeed an energy transition occurring and this will continue to gain market, policy, and technological support into the 2020s, 2030s, and 2040s, the reality is that the CO2 from aggregate human fossil fuel use through 2020 will nonetheless still be in the atmosphere and drive climate destabilization through a variety of positive feedback loops that will amplify deleterious impacts on most all planetary biomes.

2. I thank my friend Chris Farmer at Earthaven Ecovillage in North Carolina for his insights into the various distilled oil products that we are all addicted to, including the literal last dregs of the distillation barrel: asphalt, that becomes the roads upon which we drive. For a visual of most of the products distilled from crude oil, see https://www.kindpng.com/imgv/iJTwiJT_crude-oil-products-mineral-base-oil-lubricants-hd/ (accessed May 9th, 2020).

3. An economy based on "cradle to grave" material throughputs: extraction →manufacture→retail→consumption/use→disposal as "waste," often in a landfill, where some of these materials will remain for up to thousands of years. This process is highly polluting and generates huge amounts of waste, none of which factor into a "true cost" for a product. These costs are thus externalized into the bio-ecological commons, or minoritized or economically depressed communities and bodies, or are passed on to future generations. This is compared to an envisioned circular, cradle to cradle ("C2C") economy where materials are sustainably created and upcycled into new services and products in endless possible loops, and when finally worn out biodegrade, and where manufacturers and retailers have to pay the true cost of the life cycle of the product, including for restoring degraded habitats, ecosystems, and human communities (Braungart and McDonough 2002).

4. Even here, though, Amitov Ghosh shares how King Mindon and the Konbaung dynasty of Burma used oil from the fields of Yenangyaung from 1862 to 1876, prior to Drake's commercial enterprise (2016, 137).

5. Old oceans where the shells have been pressurized into liquid petroleum, and the base of "the Oil Encounter" (Ghosh 2016, 101) of the Arabian peninsula/Persian Gulf with the West.

6. Or do not, as those of us on non-tenure track contracts know all too well.

7. For rich pathways into what such scholarship may resemble, see *Energy Humanities: An Anthology* edited by Imre Szeman and Dominic Boyer (2017).

8. https://papers.aarweb.org/2018_Call_For_Proposals.pdf Accessed 21 December 2018.

9. http://www.awakethefilm.org/ Accessed December 21, 2018.

10. https://standwithstandingrock.net/ Accessed June 30, 2020.

11. https://www.youtube.com/watch?v=1Rz_TkpysKk Accessed June 30, 2020. Many youth interviewed in this ABC News feature on youth leaders at Standing Rock described the actions of ceremony and prayer at the base camp as "healing."

12. Those from various Sioux nations and other Pan-American native peoples, along with non-native peoples there in solidarity where a majority of the latter were white.

13. And just as we saw religious conservatives respond to COVID-19 in the United States by denying science and finding scapegoats, we can expect that some religious groups will steadfastly support regimes of power that violently subjugate those advocating for an equitable, just transition to post-carbon biocultures. We can also expect to see more religious dramaturgical resistance to such regimes of carbon power and petroculture, as exemplified at Standing Rock.

Chapter 5

Bodies and Religious Dramaturgy in Places of Climate Chaos

This chapter continues the pivot to more sustained case study analyses of bodily practices of religion in ecologies of place that are already being impacted by climate changes triggered by anthropogenic global warming. The trends emerging in these sites can be assumed to presage a likely future of bodily religious production, as climate changes will bring regime shifts to all biocultural practices in bio-ecological places. In short, the case studies of this chapter represent impacts on religious practice for hundreds of millions. The readily available access to data about ongoing biogeochemical shifts in these regions allows for theorizing how these shifts will impact religious embodiment in the Carboncene.

Two specific cases are useful in presenting such shifts and changes, speaking in broad terms about the religious identities and practices therein:

1. The bodily practice of Hinduism, especially of river goddess pilgrimage related to *darsan* in the Himalayas of India; and
2. The bodily practice of the *hajj* in Mecca in Saudi Arabia.

These case studies are analogous, both to each other, and to similar dramaturgical religious bodily changes that will occur elsewhere due to climate warming. The key dramaturgical practice at both locations is tethered to the performance of pilgrimage, which is currently enabled by oil. And if worst case climate scenarios unfold in the coming decades, then the dramaturgical practices in both bio-ecological locations will be radically different than they are currently. In one we may see the death of a goddess, and in the other we may see Mecca lose its geographic centrality in the practice of Islam.

In his study of pilgrimage along the holy Narmada river in India, Hindu scholar Chris Deegan writes, "Movements through sacred space require both guidelines and rules" (2000, 396). What does it mean, then, when the environmental guidelines and rules that have prescribed movements of organisms through bio-ecological spaces are in flux, such that there are major extinction events and catastrophic changes in weather patterns? Given this larger unfolding reworlding, then how will dramaturgically choreographed human movement through sacred spaces be impacted—will the rhythms, guidance, and rules for such bodily choreography inherited from elders and leaders in a tradition make sense in totally new climate regimes? Will the natural features of a place described in religious primary sources match up with new biophysical realities? Will the emic narratives, doctrines, ritual practices, and conceptions of the sacred have continued efficacy for followers of a tradition, and continue to shape bodily practices of those followers, especially if these describe or attempt to put humans in relation with sacred places that no longer exist or are no longer hospitable to human lifeways and presence?[1] These questions pertain to both case studies in this chapter, and to all religions as performed by humans, everywhere.

However, before engaging in these case studies, more discussion is needed of the specific framework of performance studies. If religion is at the end of the day embodied—somatically, aesthetically, chemically, anatomically, emotionally, psychologically, sexually, haptically, mentally—and religion is often dramaturgically performed in group settings, then an understanding of how, what, and why humans perform in a religious setting is needed. This will then let us better understand the performance of religion in the above climate changed locations, and the millions of religious bodies undertaking such performance.

UNDERSTANDING PERFORMANCE

While religious practice often follows scripts, doctrines, and inherited ritual patterns, and while it can (and often is) theatrical, it is not a form of theater. In other words, there are no directors, plots, scripts, sets, and rehearsals[2] as commonly understood when thinking of the social practice of a theater production. Rather, the performance of religious actions—ritualized, often with a script, whether individually or moreso, communally as liturgy—occur from an emic perspective in the context of "the sacred." As a scholar I recognize that the term "sacred" is problematic, as both concept and boundary marker. However, most all "insiders" who practice religion do so because of the efficacy of the practices and the doctrines of the religion they practice and subscribe to as these relate to lifeways centered around the sacred as understood by their tradition, experience, custom, and/or social pressure/network. The human datum of our performed discipline is not caught up with internal-to-religious studies Russell McCutcheon-type debates; rather, our human datum will readily accept the difference between secular theater performance, and the performance of religious behaviors. As such, as a scholar I will accept this difference in order to help bring theoretical sophistication to understanding dramaturgical bodily religious performance.

Most in the field of religious studies use the term "performance" to mean "the execution of a preexisting script for activity (as in conducting a traditional church service) or the explicitly unscripted dimensions of an activity in process (as in the spirit or quality of the service)" (Bell 1998, 205–206).[3] This understanding recognizes that human performers are active participants in creating, expressing, and perpetuating religious identities and practices. It further recognizes human performers are not passive receptacles of doctrine or practice: that as they perform there is space for ambiguity, expression, creation, and morphing of bodily religious idioms. Such understandings bleed into and are shaped by the theory of lived religion, as made popular for example by Robert Orsi and his career of work; and also by theories about religious ritual, as exemplified by Ronald Grimes and his career of work.

The performance of religion via biocultural bodies in ecologies of place means that there is a corporeal sensation and embodiment of religion that includes the semantic (the doctrines, ethics, semiotics, texts, and teachings of a religion) and the somatic (the affective, felt, embodied aspects of religion, from individual to group settings). Furthermore, these are fused in a way that can for adherents make "sense/*sense* of the unpresentable and the inarticulable" (italics in original) (Machon 2009, 20) as these relate to respective views of sacrality. Often such religious performance is to make sense of/*sense* this reality and often to interact with it or influence it in some capacity. These religious performances always dramaturgically unfold in a place, and are thus in dialogue with, and can be impacted by, the material biogeochemical flows and more-than-human bio-ecological dynamics of that place. This is so even if the intent of the performance is not to engage the bio-ecology of a place, or if the understanding of the sacred that the performance is developed around is considered to reside elsewhere or separate from the geographical and bio-ecological location of the performance. How climate changes will change the ecologies of place where religious practice always already occurs must become part of the theoretical approach to understanding religion moving forward. This is because the actual biogeophysical places where religion always already occurs, in the enfleshed *Homo sapiens* bodies that always already perform religion, are part of our metric of study. Understanding and foregrounding the agency of places in bodily practices of the religious kind means we need to see religion as occurring in "cognitive ecologies [where religion is] 'performed' in interwoven textures of action and perception" (Paavolainen 2018, 193) akin to the theorizing of Tim Ingold shared earlier in this book.

Related to this is the insight from Robert Fuller, who points out that "religious experience is, in part, due to the variety of bodily excitations" (Fuller 2008, 52). These excitations impact religious thought which itself "arises in the middle loop of brain activity whereby the input received through our bodies is translated into imaginative constructions guiding us in bodily response" (Fuller 2008, 155). This insight once again decenters and deprivileges the abstracted-out-of-physical-reality rational ego of the Human Subject and emplaces religion where it occurs: in bodies that include brains, emotions, and senses. But more so, it emplaces embodied religion in material ecologies of place, for our material bodies are always by definition in a place. To close the loop: religion is always performed in bio-ecological places that shape the embodied practices and discourses of religion—this is biocultural evolution in a nutshell. To understand this means we need to understand the material agents and biogeochemical processes that are more-than-human and how these impact our bodies and thus our thoughts and cultural practices, including those within the domain of what we call "religion." This then necessitates an understanding of climate changes, as explored below.

Furthermore, to understand religious bodies through encounters with places means that to understand religion we must consider the "ecolog[ies] of affective relations" (Simpson 2015, 72) that sympoeitically emerge in material bio-ecologies. For this reason the geographer Paul Simpson calls for a "post-foundational account of intersubjectivity" (2015, 72) that I support in this book.[4] To this I add that such intersubjectivity must extend to include nonhuman forces, materialities, and places as these interact with human animals and their embodied religious practices in a continual becoming-with. Such a view of religion as dramaturgically performed in places through bodies means we would benefit from seeing religion as occurring in "an open-ended sort of subjectivity that is perpetually emergent in its ongoing and always plural being-with bodies" (Simpson 2015, 72) of other kinds and material forces. Here we must accept the unfolding reality that climate changes as material forces will be largely maladaptive to the survival of currently evolved places and species. And we must attend to the plural being-with places within which religious bodily actions occur so we can think through how climate changes will change such bodily actions. I turn now to what this may resemble in the high altitudes of the Himalayas of India.

PILGRIMAGE DURING CLIMATE CHANGES

Hinduism in the Himalayas

I want to return to the questions posed earlier in the chapter, asked in the context of larger unfolding reworlding and the response of dramaturgically choreographed human movement through the sacred spaces this reworlding will continue to impact. These questions are poignantly apropos for any bio-cultural landscape humans co-create, and no more so than in India. One of the leading experts on Hinduism, Diana Eck, shares how, according to her own years of undertaking ethnographic fieldwork throughout the subconti-nent, the natural world for Hindus in India "*is* the religious . . . [That sacred rivers like the Ganges and others form] part of a living sacred geography that Hindus hold in common" (Eck 1996, 31; italics in original). This holding in common leads to "a sacramental natural ontology" (1996, 32) and "organic ontological vision" (1996, 32) which constitutes "a full sacred geography [such that for Hindus in India t]he living landscape is dense with signifi-cance" (1996, 33).

These insights point to a religious-based solastalgia that faces the human animals in India who claim and perform Hinduism as practice and identity. We must be leery of religious environmentalist projection, as is hinted in Ecks' characterizations of India's sacred geography, but her points remain: the landscapes of India are populated by a wide range of deities, found at temples, sacred groves, mountain peaks, at *tirthas* (sacred crossings), and encountered and seen (*darsan*) as rivers. The many thousands of sacred sites throughout India function as pilgrimage routes and anchor a dramaturgical religious economy that sees millions of devotees travel the country to receive *darsan* (sacred sight) from *murtis* (statues that house the divine) and features of the landscape that are perceived to be divine, or abodes of the divine. The import of natural features that are religiously significant for most Hindus in India are seen most iconically in the forms of rivers, beginning with the Ganges and then Yamuna Rivers. Temple complexes dwell in these natural features of the regional bio-ecology where these complexes house the divine in the form of *murtis* and that are overseen by *brahmin* priests. These various sacred sites also attract *sadhus* (holy men) and *sanyassins* (renunciates from society) whose presence help to purify and make even more sacrosanct the religious sites. Taken together, the shrines, temples, divine abodes, *tirthas*, and sacred features of the natural landscape provide continual opportuni-ties for Hindus to participate in *bhakti*. This is the devotional worship of the divine in its various forms as understood within Hindu cosmologies and resulting religious practices, occurring throughout the varied bio-ecologies of India (and also in the Hindu diaspora).

The above religious economy is dramaturgically anchored by holy sites,[5] which are the cosmological and physical foundation of a key practice in Hinduism in India: pilgrimage. At any one time there are multiple thou-sands of Hindus on pilgrimage throughout India, bringing their bodies to these sacred sites to participate in *puja* (offerings given to the divine), receive *darsan*, and interact with *brahmin* priests. These numbers swell at cosmologically auspicious times related to celestial events, birthdays of saints, or tales of miraculous events as shared in the *Puranas*. Pilgrimage is of import, but as with any human activity, it is not a politically neutral event, as "local ideas about territory, power, and authority" (Sax 1991, 147) are embedded within pilgrimage narratives and practices. These ideas relate to concepts of gender, marriage, land ownership and access, interpretations of the divine, and political back-and-forth between *brahmin* lineages and their relations with the sacred in pilgrimage spaces. The latter therefore impacts the economics of pilgrimage and livelihoods that depend on monies provided by the devout, adding a layer of political economy to India's sacred geography. And of course this religious political economy is impacted by globalization, modernization, and larger political struggles (themselves gendered [Srinivasan 2012]) of Indian villagers and city dwellers as they adapt to these forces, seen most famously in the Chipko movement (Guha 1990) and the battle over the Narmada dam (2005).

Here I will turn my attention to biocultural places of sacred significance, without which the above sacred geography would not exist: the shrines, temple complexes, and thus pilgrimage sites of the sacred rivers that flow through the subcontinent of India and which all reside at the bio-ecological origin points of said rivers. Most Hindus view rivers as divine goddesses; rivers are literally the physical, divine bodies of goddesses. Of the many rivers that are worshipped in India, two are the most sacred: *Ma Ganga* (the Ganges River), followed by *Ma Yamuna* (the Yamuna River). The origin points for these rivers are high in the Himalayas, at Yamunotri, the first village along the Yamuna, and Gangotri, the first village along the Ganges. These two sacred pilgrimage towns are accessible only in the summer months where a form of "religious hydolatry" (Haberman 2006, 45) occurs at the temple complexes, waterfalls, and streams that mark the first accessible-to-human-animals origin points of these rivers.

While these pilgrimage towns are the origin point of physical human touch with these waters the actual origin of the waters that form these sacred rivers is a mix of complex air and moisture patterns which create copious snowfall in the high altitudes of the Himalayan mountains which reside above the plains of India. As winter turns to summer the snow and ice locked in the Himalayas melts and this meltwater begins its journey to the Bay of Bengal thousands of miles away. However, on this hydrologic journey the waters pass through Yamunotri and Gangotri before merging together below Delhi in Allahabad, another sacred pilgrimage town.

The geoscientist Ellen Wohl (2011) explains how four glaciers (including the Gangotri glacier) at 18,000 feet begin to melt and as the water descends from the four melting sources the Ganges proper takes shape as a river. The snow and ice that form these glaciers come from summer monsoon rains, which provide 80 percent of the snowfall. What is unfolding in real time in the Carboncene is that these glaciers are melting faster, given rising temperatures and changing albedo (less white from snow cover, to darker from rock and soil exposure which traps more heat) in the Himalayas. They are also melting faster due to black carbon (the ashes born in air currents that originate from the many household cooking fires throughout the Himalayas). Glaciers in the Himalayas are also receiving less snowfall due to changing monsoon patterns. Furthermore, winter/spring transitional phases in the region are now "off," so that spring occurs earlier at higher elevations, bringing with it rapid snowmelt due to increased surface and atmospheric level heat. This rapid snowmelt can flood high-altitude villages. When combined with changing spring and summer rain patterns that bring higher-than-average rainfalls, this melt and rain has and will continue to wreak geological and hydrological havoc on the pilgrimage towns of the Himalayas.[6] Such havoc will not remain in the higher altitudes, but will equally impact the industrial, agricultural, and daily quotidian needs of human and other-than-human animals down glacier and downriver.

The changing waterscape of the high Himalayas within the context of global climate changes (Shrestha and Aryal 2011) will impact the dramaturgical bodily practice of Hinduism in years and especially decades to come in this ecology of place. These bio-ecological material impacts will continue, and will grow worse, possibly catastrophically so, such that rapid mobilization of localized mitigation and adaptation is required to help offset worse case scenarios (Xu et al. 2009). In terms of discourse such change is already occurring. Here the semiotics (linguistic meaning) ascribed to the glaciology and cryoscape[7] in this region as it rapidly changes (Nüsser and Baghel 2015) is shifting. This shift in discourse signals at changes in embodied practices of labeling, seeing, knowing and thus conceiving of changing water/weather patterns, especially when compared to the lived past of locals in the region. This knowing about weather patterns, and the language used to describe the glacial landscape based on lived experience is a form of embodied environmental knowledge unique to living in high Himalayan regions of Hindu India, formed in ways described by Tim Ingold earlier in this book. Such embodied environmental biocultural knowing will be grafted onto and woven into existing Hindu biocultural lifeways in the bio-ecologies of high Himalayan places. Given the symbolic significance of Himalayan glaciers to Hinduism, as origin points of divine riverine bodies, understanding the response to their melt becomes a focus of religious studies. As Nüsser and Baghel explain, "glaciers do not *just* melt; they are imbued with cultural, scientific, political, and aesthetic meanings" (2015, 150). We must add here, too, that in these bio-ecological places they are thoroughly imbued with religious meanings. As scholars of religion we must attend to how bodily interactions with glacial melt as mediated through Hindu cosmologies, including the physical act of performed pilgrimage and the semiotics thereof, signal the likely onset of the performance of new types of Hindu dramaturgical bodily religious practice.

Given the above sympoietically emerging bio-ecological and biocultural context I can here only turn to thought exercises of predicted changes in bodily practices based on regional climate models. The reality is that more "real-time" fieldwork and ethnography is needed to tease out the specifics of what I can only predict here. This reality, however, points to the important insight that we are not undertaking good religious studies scholarship, especially of ethnography, if we are not attending to rapidly changing climate conditions and how these are already impacting the dramaturgical bodily performance of religion.[8] We know that this has happened in the past—whole biocultures have adapted to changing bio-ecological conditions. It is myopically naive to assume humans today, saturated as we are in petrocultures, are immune from these same materially agentive bio-ecological dynamics. If "bodily enculturation" is about how material cultures limit or help promote the capacities of religious production through bodies (Carp 2014, 478), then understanding how global warming is changing the material culture (especially the bio-ecological aspect thereof) of the Himalayas is required. Only by factoring this in can we engage in a religious studiy that needs to be attuned to the material realities of the Anthropocene epoch. Such bodily enculturation is everywhere, of course, with the same growing limits and choke points of embodied cultural production in a world of growing biogeochemical and bio-ecological flux; therefore this need is not just germane to understanding Hinduism in the Himalayas.

One of the religious studies scholars who have spent a majority of their career considering environmental changes and their impacts on religious practice is David Haberman. His fieldwork throughout India expertly weaves together textual analysis, ethnography, and the environmental sciences, all in an effort to better understand how environmental changes are impacting Hinduism. His magisterial *River of Love in an Age of Pollution* investigates the pollution of the Yamuna River and how this pollution is impacting the theology of the river. Haberman's book includes ethnography undertaken along almost the entire length of the river, including time spent at Yamunotri. As he describes,

> Today, many people visit and honor Yamuna at her origin. Thousands of pilgrims make the long trek up the steep slopes of the upper Yamuna valley to visit this natural shrine. Most pilgrims do so as part of a greater pilgrimage known as the Char Dham Yatra, the "Pilgrimage to the Four Sacred Abodes." During my four stays at Yamunotri, I met people from every major state of northern

India but never encountered anyone performing solely the pilgrimage to the Yamunotri. All intended to visit the three other sacred sites of the Char Dham pilgrimage, one of the most important pilgrimages for Hindus of northern India. Many whom I interviewed told me that as Hindus they believe they are to perform this pilgrimage at least once in their lifetime. Many identified it as the most important one they would ever undertake. The four sites of the Char Dham Yatra are, in order, Yamunotri, Gangotri, Kedarnath, and Badrinath. All these sites are associated with the Ganges, since they are all located near the sources of rivers that eventually feed the Ganges. (2006, 47)

There is no succinct formula for describing the variety of embodied dramaturgical activities that occur during a Hindu pilgrimage, let alone to high in the Himalayas. However, the following bodily needs and movements are necessary conditions for any devotee involved with pilgrimage to the origin towns of the Ganges and Yamuna:

1. Travelling to and from Gangotri or Yamunotri.
2. Locating places to sleep and eat while at either pilgrimage town.
3. Bathing, in both secular form and also as a form of purification and *darsan* with the liquid material divine of either *Ma Ganga* or *Ma Yamuna*.
4. Locating and bringing *puja* offerings to the river goddesses, where such *puja* will involve the recitation of chants, hymns, and prayers, often overseen and aided by *brahmin* priests.
5. On important feast or worship days the circumambulation of the *murtis* around the village.
6. Walking along narrow trails, up hills, across bridges, through temple complexes, along the rivers, and to waterfalls. Often this is done with a group and aided by *brahmins*, but such activities can also be done alone as a form of meditation and *darsan*.
7. Consuming *prasad*, food blessed and sanctified by the divine via the aid of *brahmins*.
8. Caring for *brahmin* priests who live in the temples and must be fed and undergo other daily quotidian needs, and who themselves take care of the *murtis* in their care and who keep the temple complexes purified by performing various rituals.
9. Having enough food and water to keep one's body full and fit so that walking and hiking can occur.

This entire list contains bodily practices. Many are religious-specific dramaturgical performances undertaken by bodies, typically within religious performance spaces and group settings. Some are bodily practices that are preconditions for engaging in pilgrimage, and without them there would be no religious-specific embodied practices. Also notice all nine items of pilgrimage at contemporary scale are utterly dependent upon fossil fuels—from the transport of *puja* items, to the transport of human bodies up into the Himalayas, to the transport of food needed to eat so that people can engage in the ritual worship of *Ma Ganga* and *Ma Yamuna* at these origin temple complexes. This means the material agency of old organisms as combusted through the pistons of vehicular engines (plane, train, auto, rickshaw, bus—themselves reworked materials of the earth) creates the ability of human bodies to have religious agency. It also means that as we descend Peak Oil and it becomes harder to pay for and/or locate oil that the above dependencies will make the bodily practice of pilgrimage to the high Himalayas harder to perform. It also suggests shifts in the to-date stability of weather patterns and geological formations of the region where this stability has been required as a material context for the embodied performance of human actions related to these nine categories of behavioral action. Yet we know that this stability has never been, and for sure what quasi-stability there has been over the last few thousand years that Hindus have already dramaturgically adapted to is now over. The cascading bio-ecological shifts from climate warming are here and will materially shift the entire region and reform its ecology of place, and thus the embodied and performed religious dramaturgies within that material place.

Furthermore, how the goddesses function in the lives of devotees and within Hindu theology, broadly, may not be able to survive the hyperobject of climate change. Both goddesses are believed to wash away impurities and sins, and are seen as loving protectors and caregivers, capable of removing the devotee's *atman* (innermost divine soul) from the wheel of rebirth, where this soul instead will reside in bliss for eternity at the foot of the goddess in question. Devotional rituals that are always practiced and performed by bodies in relation to these waters may not survive climate change. For example, in 50–100 years there may be minimal glacial melt, meaning certain areas of ritual bathing may no longer exist; or current temple structures may have to be relocated to be near much lower flows of water; or at certain times of the year there may be no water at all flowing! In short, the embodied dramaturgical rituals that bioculturally evolve in this ecology of place may be radically different than practiced on a planet that was under 350 ppm of CO_2 and had less methane and water vapor in the atmosphere. That planet had material Holocene weather patterns, including monsoon rains, snowfall patterns, and glacial melt around which the embodied and performed practices of devotional pilgrimage developed at Yamunotri and Gangotri. On a post-Holocene climate changed planet the vulnerabilities of certain geographical regions will be more pronounced than others, including in the Himalayas. As the historian Vinay Lal explains, "The ecological inheritance of Hinduism is . . . laden with ambivalence. Climate change may perhaps bode greater ill for Hindus than for adherents of other faiths [*sic*] for none other than the reason that so many of Hinduism's principal pilgrimage sites are associated with rivers, mountains, or other manifestations of nature" (Lal 2015, 398), all of which will be impacted by a changing climate.

On our Anthropocene home of >415 ppm CO_2, with CO_2 very likely rising to 500 ppm CO_2 and beyond, the material biogeochemical processes of the earth are changing. This means that in the Himalayas the monsoon is coming at different times and with different frequencies, and glaciers are melting faster and faster. What happens when there is less and less spring melt? What happens when there are deluges of extreme glacial and snowmelt that may wash away the bridges, trails, and temple complexes of Yamunotri and Gangotri? What if these environs become too dangerous to inhabit year around as is currently done by a core group of *brahmins*, and therefore the *murtis* and temples are unable to be cared for and the goddesses may no longer be present as they may choose to abandon the temples and *murtis* where they are believed to reside? According to orthodoxy, if this were to occur it would be a sign of the goddess withdrawing her presence, which would radically alter the biggest form of river goddess worship in all of Hinduism. How can you have pilgrimage and worship of a goddess who is no longer present? If many Hindus on pilgrimage report affective bodily responses to moving through and in spaces conceived of as sacred such as "'magical' and 'sacred' and 'holy' to describe both the place itself and their experiences of it [and of] being transformed by the ambience of the location" (Fuller 2011, 70), will we see similar such bodily affects in the future? Or will the future affective responses be ones of loss, fear, dismay, and abandonment?

If the headwaters change location, or dry out, and the temple complexes become abandoned, will there be an intergenerational religious forgetting? What will elders say, who remember (think here of solastalgia) a vibrant pilgrimage experience, complete with stable climate, in the 1990s or 2000s, when they revisit Yamunotri and Gangotri with grandchildren in the 2030s, 2040s, and 2050s? Will there even be a fully functional Ganges River in the 2050s, given predicted loss of glacial ice in the Himalayas? And if pilgrimage deals with "questions concerning informants' views and constructions of locality, landscape, mobility, space, place, the national and the transnational" (Coleman and Eade 2004, 6), then how will these categories shift for informants as they struggle to navigate climate triage (collapsed economies, transit systems in shutdown due to lack of fossil fuels, food shortages, changed weather patterns that include record heat in the plains of India) in order to access these pilgrimage towns? And if certain rituals are geared toward controlling the climate and such perceived control is no longer evident, what happens to the logic behind such rituals and the social power of those performing them (Halperin 2017, 70–71)? These are open-ended questions, but they need to be asked and answered, and to date most scholars in our field are not asking them, or at least not asking them consistently enough, with the interdisciplinary expertise they imply.[9] In order to better understand the bodily production of Hinduism on a climate changed planet, these are precisely the types of questions we need to be asking, and to date too few are asking.

Yet in asking them we must recognize that while these questions are important for better scholarship in a materialist context, the training we bring to the questions may not reflect the lived experiences (yet) of interlocutors. This was the experience of Georgina Drew in her fieldwork at Uttarkashi, a pilgrimage town on the Ganges devoted to Lord Shiva. While there she "was surprised to find that the estimates for glacial retreat and the scientific discourses on climate change were not yet topics about which many of the people I spoke with expressed extensive familiarity. My conversations with people at Manikarnika ghat were particularly indicative that not everyone was aware of climate change or that, at the very least, they did not perceive it or fear for its effects in the same way I was trained to" (Drew 2012, 349; see also Byg and Salick 2009; Halperin 2017).[10] So while we must incorporate climate science and climate scenarios into our understanding of religion-in-action through bodies-in-places, such incorporation must also attend to the lived experiences of those who perform and embody religion in the places being changed by climate. They are already aware of such changes, whether explicitly (as at Standing Rock) or implicitly (see next chapter), and their bodily practices are changing. The likely impacts of this climate changing future is made even more evident by the changes pilgrims on the *hajj* will in and around Mecca in Saudi Arabia.

Islam in Mecca

This section moves from the high elevation and current abundance of water in its various forms of the Himalayas, to low elevation and a bioregion that has for centuries had very low levels of water available to human bodies, Saudi Arabia. The United Nations recognizes Saudi Arabia as a water-scarce country, meaning less than 500 cubic meters of water is available per capita per year. This water scarcity status is a result of there being no large rivers or lakes on the Arabian Peninsula, coupled with high rates of groundwater depletion over the last several decades from Saudi Arabia's agricultural sector. Erica DeNicola, et al. explain this, and Saudi Arabia's overall geography of water, by writing: "Located in the tropical and subtropical desert region of the Middle East between the Persian Gulf and the Red Sea, temperatures can reach more than 50 C (122 F) in some areas, producing overwhelmingly hot and dry conditions. Long-term average rainfall across the country is 114 mm per year . . . Saudi Arabia is one of the poorest nations in terms of natural renewable water resources" (2015, 343).

This water scarcity is just one of the brittle aspects of Saudi Arabia's current oil-based economy and its political regime rooted in the Salafi version of the Hanbali school of Sunni Islam. Such scarcity, which threatens long-term economic, political, agricultural, biological, and public health stability, despite oil monies for desalination and reused wastewater interventions, will become increasingly exacerbated under climate change (Sowers et al. 2011; Williams et al. 2012). A study of 21 temperature stations throughout the Arabian Peninsula (including six in Saudi Arabia) that analyzed observed temperatures over 1980–2008 found a very clear trend of already occurring warming that is statistically significant (AlSarmi and Washington 2011; see also Almazroui et al. 2012 who had similar findings). In short, human-induced global warming is already impacting the Arabian Peninsula, with some of the sites reporting an average of 1.5 °C, up to even 2.27 °C, increase in warming over a decade compared to baseline, representing a warming trend that is 1.5–3.5 times higher than mean trends around the rest of the planet (AlSarmi and Washington 2011, 12). Overall, the prediction is that by 2050 there will be an increase of temperature across Saudi Arabia of 1.8–4.1 °C (3.24–7.38 °F) (Chowdhury and Al-Zahrani 2013, 1968; similar findings were achieved by Almazroui [2013] who calls for more rainfall near the Red Sea but less near Mecca in the southeast, and a minimal warming of .65 °C per decade between 2021–2070). This is alarming, as current summer highs already reach a sustained 45 °C (113 F), so it is possible that by 2050 sustained daily temperatures in the summer will be 49 °C/120 °F, with even higher occasional heatwaves (at the time of this writing, record heat waves have already approached 130 °F in the region). The irony of the above impacts of climate change on Saudi Arabia is that the Saudi government historically is one of the biggest global players in obstructing meaningful action on mitigating global warming (Depledge 2008), while the government benefits fiscally from global demands for carbon in the form of oil (but does so due to violence and undemocratic policies [Mitchell 2011, 253]). A further irony results from the overall annual aggregate carbon footprint of those participating in and hosting the *hajj*, one of the most significant rituals in Islam.

Islam is a religion that can be understood as being built around five key pillars of practice.[11] These are *salat*, or daily prayers; *sawm*, or fasting during the month of Ramadan; *zakat*, or participating in charitable acts, typically of financial donations to the needy; the *shahadah*, or testimony of faith in Allah, the Muslim monotheistic god, and lastly, and of importance to this chapter, the *hajj*.

Muslims[12] trace their understanding of the sacred, and thus religious identities, back through a series of prophets, where Muhammad is known as the
"seal of the prophets," who received Allah's ("the God") final revelation
through the Angel Gabriel as articulated to Muhammad by Gabriel. This
revelation was eventually written down as the *Qur'an*. It is believed that
new revelations given to Muhammad that contradict or embellish revelations
given to all prior prophets take precedence, otherwise the messages given
by Allah to prior prophets in Christian and Jewish traditions hold as valid.
Muslims trace humanity back to the original Biblical Adam and Eve of the
Hebrew Bible's Genesis creation story and accept that God entered into a
covenant with Abraham and his descendants to be his chosen people.

According to Genesis, Abraham and his wife Sara had a hard time conceiving a child. In the meantime, Abraham had a son, Ishmael, with his
concubine, Hagar. After this Sara became impregnated with this leading
to the birth of Isaac, who God through an angel directed Abraham to sacrifice. Muslims, however, believe that it was Abraham's son, Ishmael, born
of Hagar, who was to be sacrificed. Furthermore, they believe that out of
jealousy Sara demanded Abraham send Hagar and Ishmael away from their
compound. It is this story that anchors the *hajj*, for Hagar and Ishmael end up
in what is Mecca, and it is prophesied that the true message of God would in
the future be made available to all humans through a final prophet in Mecca.
In short, Hagar and Ishmael kept God's pure lineage alive in Mecca, to be
redeemed after the Jews and then Christians who would have valid prophets
but not live up to the teachings and covenants these prophets brought to their
respective communities. In the meantime, the monotheism brought by Hagar
and Ishmael to Mecca and the *Kaaba* (black box, in the center of which sits
a meteor believed to be seen by Abraham and Ishmael, given to Abraham by
an angel as a cornerstone in the first mosque believed to be built by Abraham)
was forgotten, with Mecca falling into polytheism and *djinn* worship. It is this
context that Muhammad was born into and through a series of miraculous
and military events became de facto political and religious leader of Mecca
and then the Arabian Peninsula over the period of 620–632 CE, to which he
brought once again the pure monotheism of Allah and Allah's teachings.

The five pillars are built upon the above sacred story. For example,
Muhammad prayed five times a day, so all Muslims do. At first he prayed
toward Jerusalem, but then turned the direction toward Mecca, thus making
a break from the holy city of Jerusalem and making Mecca the holiest city
of Islam. Specifically Muslims all pray toward the *Kaaba* in Mecca, making
it an archetypal form of Mircea Eliade's *axis mundi*. Muhammad would fast
during Ramadan, and it is during this time that he received Allah's revelations
through Gabriel. Hence, all Muslims are expected to fast (no food, drink,
smoking, or sex from sunrise to sunset) during Ramadan. The *shahadah* is a

reaffirmation of monotheism, said over all Muslims when born, or three times with sincerity if one converts to Islam. *Zakat* reflects Muhammad's position as a social reformer in Mecca, where he advocated for the well-being of widows, orphans, and the poor. And lastly is the pillar of the *hajj*—Muhammad removed the residues of polytheism from the *Kaaba* and made it once again the center of Islam and its practice. The performance of a regional pilgrimage to Mecca and the *Kaaba* already existed but Muhammad turned this into a pilgrimage to the *Kaaba* in honor of Allah.

The *hajj* is an obligated pilgrimage all Muslims must take to Mecca unless they are physically or financially unable to do so (in which case it is accepted practice that someone else can go in their stead). As there are over 1 billion Muslims[13] this means that in any given year there are millions of Muslims in transit to and from Mecca, the physical site of the *hajj*. Because the Muslim calendar is a lunar calendar that begins with the *hijra*, or the Prophet Muhammad's emigration to Medina in 622 CE to escape persecution in Mecca, the *hajj* is often held in the late summer and early autumn months. This means that in these lunar years it falls at the tail end of the hottest summer months—months that will get warmer by 3–7 °F in the coming decades.

The dramaturgy of the *hajj* itself occurs over multiple days and is highly regulated, with both male and female attendees wearing white sheets, or *ihram*, to recognize entering the sacred space of Mecca as pilgrims. Once in sacred garb and bound by standards of behavior (for example, no sex during the pilgrimage, and bathing only with pure water and no soap), pilgrims begin by circumambulating the *Kaaba* counterclockwise seven times. There are so many pilgrims in Mecca at this time, circling the *Kaaba* that they are visible in satellite photos taken from space. From the *Kaaba* pilgrims then undertake *sa'ī*, or "exertion," which consists of rapidly walking between the hills of Safa and Marwa. This re-enacts the plight of Hagar running back-and-forth between these hills, looking for water for herself and Ishmael after being abandoned by Abraham. Hagar found water, thus saving Ishmael and herself, and Muslims believe this water still pours forth at the well of Zamzam, where this well was revealed to Hagar by God himself. The next key stage of the pilgrimage takes place the following day when everyone moves to Mina and engages in focused prayer. The next stage is to move nine miles away to the plain of Arafat where all pilgrims camp. In today's *hajj* economy the wealthier camp in air-conditioned tents, while others sleep under the stars. At Arafat pilgrims stand from noon until the sun sets, speaking Allah's various names and asking for his forgiveness and blessings. This is followed by walking from Arafat to Muzdalifa for evening prayers and another night camping. Upon waking pilgrims decamp to Mina and engage in a ritual "stoning" of Satan, where pilgrims throw pebbles at three small pillars. From here pilgrims spend the rest of the *hajj* between Mina and Mecca, specifically the *Kaaba*, until the *hajj* is officially over.

The above ritualized actions that constitute the *hajj* are central to Muslim belief, identity, and practice as "The hajj, like congregational prayer, is a highly public event, and the honorific title of *hājī*, which those who have completed the hajj confer upon themselves, instantly indicates fulfillment of one of the most important obligations in the Islamic canon" (Turner 2011, 163). The inscription on bodies, and performance of religion via bodies, throughout the *hajj* is prolific: from entering the sacred space of Mecca (allowed only for Muslims), to changing into ceremonial dress and undertaking cleansing ablutions, to circling the *Kaaba*, to participating in prayer and *salat*, to various periods of fasting, to drinking holy water, to throwing stones at pillars representing Satan, to standing in the sun as it sets with prayer and attention focused on Allah—these combine to create "modalities of agency" (Mahmood 2005, x) in those present. Yet these modalities of agency are within materialist settings that move beyond the human, and that require multiple uses of fossil fuels at contemporary scales to make happen:

1. Transport to and from Mecca. For most pilgrims the world over this is done by vehicle (car or bus), and now often includes air travel, as well as motorized boat for some.
2. Staying in Mecca at hotels, which are powered by fossil fuels (especially their air conditioning units).
3. Eating multiple meals per day over seven–nine days—this food is not being grown in Mecca so is shipped, bringing with it hidden water and hidden fossil fuels (natural gas for fertilizer, oil to ship and refrigerate, etc.), and then is sold in stores using more fossil fuels.
4. Transport to various pilgrimage sites via bus or auto.
5. For some, fossil fuels to power the air conditioners hooked up to tents.
6. Fossil fuels to transport the workers who help run the *hajj*, and the food they need to eat.
7. Fossil fuels to power Mecca as a city and that allow city operators to provide various services to pilgrims (law enforcement, electricity, sewage, water treatment).
8. Growing the cotton (typically by fossil fuels) that become the *ihram*, and then powering the plants where the millions of *ihram* are made, and then shipping the *ihram* to Mecca (or wherever around the world pilgrims buy this sacred garb in preparation for the *hajj*).

Here, then, is the irony mentioned earlier—the material agency of fossilized bodies of sea organisms and plants that are now burned to power the *hajj* is contributing to the rapid warming of Mecca; warming that will only require further uses of fossil fuels to keep pilgrims cool and safe so the *hajj* can continue. And just like with pilgrimage in the Himalayas, the entire *hajj* consists of bodily practices that are utterly dependent at this point upon fossil fuel petroculture.

It is hard to envision a transition away from[14] using fossil fuels in those participating in and hosting the *hajj* such that for the foreseeable future the religious bodily practice of this required part of Islam is dependent upon old forms of carbon. It's also hard, given the climate and temperature predictions for the region, to see how ecodisruption does not make Mecca nearly uninhabitable in summer months at some point in the coming decades. If this catastrophic warming is the future for this bio-ecological place, then it is highly probable that the *hajj* as currently known and practiced will either cease to exist, or shift entirely to winter months, or be relocated. Whole pilgrimage-based industries may collapse, while whole new industries may develop (air-conditioned tents in Mecca). And should any rigorous economic or climate policies develop that limit air travel due to capping aggregate global CO2 emissions (not at all likely at the time of writing), this may drastically alter the number of pilgrims journeying to Mecca per annum. However, a return to old pilgrimage forms of travel may emerge: boat, animal, vehicular, and foot travel, as the faithful used to use before the advent of the combustion engine. Yet this is unlikely, as climate models suggest the region will be too hot for healthy habitation, and even if people can go there, they may not be able to stay for more than a few days. We do not know what it will take to sustain a metropolis when the daily heat for months on end is consistently over 120 or 130 °F. The need to restructure the built environment of Mecca to adapt to climate change may be beyond the threshold of human ingenuity, let alone the earth's ability to provide the resources needed to transition Mecca.

Like the forecast of minimal snowmelt, and that coming in a rapid burst, in the Himalayas 50–100 or so years out, the forecast for Mecca is as bleak and tragic. It may be that the religious significance of these places will symbolically remain, but the biogeochemical and bio-ecological context of them will have altered so much as they are either unable to be visited; or will not be visited at the present numbers; or may not be visited for as long, suggesting much-shortened *hajj* and glacial *darsan* dramaturgies. What this means to the emic understanding of both religions (and all religions), and moreso for bodily practices, is something scholars will have to remain attuned to and track in our respective research regimes.

WHERE WILL (RELIGIOUS) BODIES GO?

The religious identities and practices tethered to the above biocultural case studies at these two bio-ecological locations account for approximately one out of every five human beings on the planet. The pilgrimage sites of the

Himalayas, specifically of the Ganges and Yamuna rivers, are visited by millions, and are sacred to hundreds of millions. Mecca, site of the *hajj*, is a sacred site to over one billion, who in theory must visit Mecca at some point in their lifetime. It is not an understatement to say that as these two bio-ecological locations are impacted even further by climate changes that the global landscape of religion may look very different in 20–50 years, depending on which climate models one chooses to follow.

It is hoped that the brief exploration of sacred sites and religious bodily practices at the two geographic ecologies of place offered in this chapter helps to articulate why and how we must attend to climate changes in our understanding of religious bodies. The forecasts for Mecca and the Himalayas are sobering, or even depressing and angering, as models predict catastrophic stresses on the respective bioregions and all organisms therein. Political will exerted at any time from the 1970s onwards would have helped "save" these regions where that will would need to have served to limit the extraction and use of fossil fuels. Sadly, such political will was thoroughly lacking (and is still lacking), given the structure and power dynamics of "carbon democracy." There is nothing positive or celebratory about the possibility of losing these sacred sites, and the religious bodily practices that have evolved at them over hundreds to thousands of years, to say nothing of the other flora and fauna threatened in these places.

However, bodies can be resilient. This means religious bodily production can also be resilient, and may help larger ecosystems be resilient as humans-in-bodily-places materially adapt to climate changes. The next two chapters join the earlier #NODAPL case study and foreground where such religious bodily practice is attempting to cultivate climate resiliency at the interface of religion, politics, ethics, doctrine, and prosustainable bodily engagement. It builds on the insight offered by Willis Jenkins, Evan Berry, and Luke Beck Kreider who write, "Religious engagement with climate change may ... happen outside of conventionally religious communities (e.g., the Catholic Church) and beyond formal reflection on discrete traditions (e.g., Buddhist responses). Religious interpretation of climate change may also happen within secular social movements, and in forms of nature-based spiritualities that draw on multiple, hybrid streams of culture" (2018, 100). It is to three such forms of nature-based, hybrid religious thought and practice in an era of climate changes, and one Buddhist, to which I turn.

NOTES

1. A nightmare scenario that I do not have time to explore in this book, but that is within the realm of statistical probability given climate change models, is the meltdown

of nuclear reactors in coastal lying regions the world over. Sea level rise will prove such reactors to be very brittle and as they potentially meltdown due to submergence, the resulting release of radiation (think material agency) may make their surrounding environs uninhabitable by human animals. If full meltdown does occur then it is very likely that all nearby religious sites and the bodily practices associated with them will cease to be visited, interacted with, and performed. What does this mean for those involved in such religious traditions at these sites that may become irradiated?

2. This is not to say these do not exist in religious performance. Clearly, they do, but in a context of religious performance, not secular theatrical performance.

3. This leaves aside the reality of improvisation, which is also part of performance. I thank my editor Richard Carp for pointing this out.

4. This is a move beyond the assumed isolated autonomous Subject that is the foundation of humanism, as well as modernity. A post-foundationalist move "entails an emphasis on material relations between bodies (and the world more generally) which 'selve' in their singular-plural being-with other amid the shared horizon of the sense of the world. These bodies are not to be conceived of as being governed by a Subject which empathically recognizes the other and communes with them, nor a subject called into being by an absolute Other. This originary sociality does not mean that there is a shared origin or unity in community as there is no origin or ground to be shared. Rather, it is a body that is productive of subjectivity in its exposure, in its being-*with* other bodies. Secondly, this logic of the being-with of bodies and world entails an open-ended sort of subjectivity that is perpetually emergent in its ongoing and always plural being-with bodies. Bodies and selves are always already dis-posed, in co-motion" (italics in original) (Simpson 2015, 72).

5. My word choice of "anchored" here does not mean that there is no fluidity, change, or ongoing revelatory experience at existing or even new holy sites. Rather, it is chosen to signify the foundational import of these sites to the bodily dramaturgy of pilgrimage. I intend the word the same way an anchor functions in holding a ship in a place. That is, these sites literally form an anchor that holds in place (again, this does not mean no adaptive fluidity or emergent change in biocultural practice) the larger constellation of pilgrimage-based dramaturgical practices that revolve and evolve around them.

6. Seen most tragically in 2010 in Ladakh, which triggered regionally specific Buddhist responses, with this informing some of the adaptive responses covered in chapter 6.

7. "This neologism is derived from the *cryosphere* as an object of scientific study, glaciers as a *living landscape* (as opposed to wilderness), and finally from the emergence of Himalayan glaciers as a dynamic, globally imagined *mediascape* in the sense of Appadurai" (italics in original) (Nüsser and Baghel 2015, 150).

8. For an example of what climate-informed ethnography should look like in this bioregion, see the work of anthropologist Georgina Drew (2012)

9. For an example of the type of climate change religious studies scholarship that takes these future scenarios seriously in its theorizing, see Jace Weaver (2015).

10. One area of potentially rich research is if scholars were to undertake longitudinal studies of either the same geographic locations and how religious actors are

responding to climate changes in those locations; or of the same religious actors and their responses to and conceptions of climate changes over a series of ten to twenty to thirty years.

11. The five pillars of practice is often a heuristic historically used by Western scholars to help categorize and define Islam.

12. The next few paragraphs provide an overview of basic Islamic history to provide context for understanding the *hajj*. Those readers familiar with Islam are invited to skip this section.

13. Sunni and Shia are the two main types of Islam, with Sunni and its four key legal schools accounting for approximately 70 to 80 percent of global Muslims, and then a variety of Shia subtraditions.

14. Clearly it is possible such a transition could occur, as the *hajj* predates the rise of fossil fuel use and modern petrocultures. Currently, however, it is hard to imagine those on *hajj* taking months off work to travel by sail, train, bicycle, animal-power, and other forms of travel less dependent on fossil fuels than planes and vehicles.

Chapter 6

Regenerative Thrivability and Flourishing—Ladakhi Buddhism in the Age of Climate Change: Constructing Identities and Adaptive Responses

In her *A Field Guide to Climate Anxiety* environmental humanities scholar Sarah Jaquette Ray points out that the "anticipatory grief" (2020, 21, quoting of original phrase from Amy Spark) of climate change can lead to a sense of fatalism and inaction. This is because the scale of climate change is so huge and the implications so daunting. She shares that by focusing on and helping to generate climate solutions those engaged in adapting to climate change are psychologically and affectively better prepared to continue with such solutions-oriented engagement. This allows people to continue helping to create the just, equitable, post-carbon (Heinberg 2010) biocultural lifeways that are needed for the rest of our species' time on the planet. Such concrete, material actions signal a move beyond sustainability to active and hopeful thrivability (Edwards 2010). Can religious practitioners through performative, bodily practices contribute to such thrivability? This and the next chapter work through possible answers to this question. Calling back to the insight from before that hegemonic oil lifeways are dependent on and promote disparate forms of violence and that creaturely solidarity is needed to navigate them, these case studies indicate that thrivability and resilience as generated via human bodies in ecologies of place can become forms of such creaturely solidarity.

This and the next chapter also work through these answers in part through the lens of regenerative sustainability. How humans use and interact with bio-geochemical flows and the products of thermodynamics matter, ethically and physically. Literally—here again the feminist materialist lens I have adopted as part of this book returns, where I share this movement's goal of "investment in *becoming-with-context, situated knowledges* and *speculative alter-worlding*" (Asberg et al. 2015, 164, italics in original). Human lifeways are not static and are made possible only by and through the becoming-with-context of trophic flows of energy. This hard biological materialism recognizes that we are animals within larger complex systems that are adaptive, hierarchical, and multi-scalar; that as animals with culture, where especially place-based religion can be a form of situated knowledge central to such culture, we self-organize within a "small set of critical self-organized variables" (Holling 2001, 391) that allow us to evolve bioculturally in ecologies of place. We are constituted by smaller microsystems, and at the same time reside within larger systems that are part of even larger scaled systems, all defined by "increasing complication" (Anderson 1972, 396). However, because these larger systems are shaped by emergent properties (including radical loss of complexity brought about by human actions, like habitat fragmentation or extinctions; or overloading the atmospheric commons with a few greenhouse gases) and can undergo stochastic, tipping point shifts into new regimes (including radically impoverished ones that over evolutionary time will become complex again), it is bioculturally adaptive to have flexibility in our ideas (Bauman 2017) and behaviors. This flexibility can be imagined through speculative alter-worlding. We already see this flexibility through the trope of lived religion; for example in flexible, adaptive responses to perceived biosocial damages caused by industrial agriculture. Here religious practitioners within Judaism and Christianity have turned theocentric values into place-based sustainable behaviors related to agriculture (LeVasseur 2017). Does the same process exist with larger movements related to climate changes and living within the nine planetary boundaries recognized by the Stockholm Resilience Centre?[1]

The following case, situated in the Himalayas of Ladakh, provides a bridge from the analyses of Gangotri and Mecca last chapter, into the proresilient religious dramaturgical embodiment and performance of this and the next chapter. This case study also presents animist sensibilities, but not in the ways of the next chapter's Global North posthumanist case studies. For those in Ladakh, other-than-human agencies are chthonic and damaging beings and forces that need to be actively pacified and contained by Buddhist monks via the monks' embodiment of the dharma. That these forces are seen as being responsible for recent catastrophic floods and shifting weather patterns in the Ladakhi bioregion presents a different ontological understanding of climate changes. However, it is one that is increasingly being challenged by emerging scientific discourses about climate change entering into the region, including into monastic communities.

In this chapter we will analytically and geographically travel to Ladakh, India, where in the summer of 2019 I taught a study away course on religion, sustainable development, and climate change, and my colleague at the College of Charleston Zeff Bjerken taught a course on religion and globalization. Our home base was the capital city of Leh, one of the fastest-growing regions in all of India, helped along by the popularity of the 2009 Bollywood hit *3 Idiots* and the tourist gaze made possible by the expendable income (and fossil fuels accessible to) of India's rapidly growing middle class. However, this growth is not without its own forms of cultural and ecological violence and traumas, fueling regional tensions along permeable lines of: Muslim/ Buddhist, tourist/local, foreigner/local, affluent/poor (complicated and amplified by India's cultural divides around caste), and embrace of growth/critical of growth.[2]

Ladakh is known as "Little Tibet," for its long history of both Tibetan Buddhism and tantric Buddhism. It is also a high-altitude desert, receiving approximately four inches of rainfall a year, but (currently) abundant snow. This snow melts in the summer, providing irrigation to fields of barley and peas, while helping replenish glacial lakes and aquifers of fossil waters. The below situates these confluences, reflecting on how Ladakh, informed by Buddhist practices, is responding to directly experienced and observed (and in geographical timescales, rapidly) changing biogeochemical cycles, bio-ecosystems, and ecology of place. It must also be acknowledged that there are multiple stakeholders in this case, such as traditional ecosystem peoples in Ladakh whose lifeways are threatened by climate change; omnivores in Ladakh, and especially Leh, who are facing disruptive effects of climate change; and ecological refugees (becoming the urban poor of Leh) who are also being impacted by climate changes (see below discussion of Gadgil and Guha on these three groups). All three groups are also impacted by larger regional trends related to climate change, tourism, the militarized border with Pakistan, and globalization. While the latter two groups (omnivores and ecological refugees) are changing the bio-ecological dynamics of the region, they have not historically been involved with developing and maintaining the Buddhist-influenced agrarian biocultural practices and identities of the region.

Engaging in this case, it is necessary to also be cautious of a long-standing history of romanticizing the Himalayas, and especially its Tibetan Buddhist regions, as a Shangri-La (Lopez 1998) where indigenous Buddhists have always practiced a fully sustainable and enlightened life until the onset of tourism. Although well-meaning, Helena Norberg-Hodge's best-selling *Ancient Futures* (2009 [1991]) about Leh and Ladakh being an ecotopia prior to Western tourism is a prime example of such narratives.[3] Such romantic notions belie that humans everywhere, including in Asia, often presented as the foil for an earth-destroying West, have always impacted their environments, and often toward impoverishment (Tuan 1970). Situated on the historic Silk Road, Leh has a long history of trade. The creation of agricultural hamlets fed by snowmelt and monasteries on tops of plateaus (from which to pacify and conquer local spirits and demonic forces, while showing the power of the *sangha* [institutionalized monastic community and their lay followers] over the regional feudal monarchy) has directly shaped and impacted the geography and ecosystems of the region.

That said, it is very clear that a "traditional" Ladakhian biocultural lifeway (the ecosystem peoples of Gadgil and Guha discussed below) was (and where it currently exists in full or part, is) much more sustainable than any form of petroculture. Ladakhian traditional ecological knowledge informed and helped to create an almost circular economy that generated minimal "externalities," largely because for thousands of years their economy had never "advanced" past bronze and iron age technologies (i.e., a la Wes Jackson, Ladakh was historically shaped only by the first three pools of carbon). This biocultural lifeway was based on a syncretism of Tibetan Buddhist and folk religion, an agrarian seasonal economy that dictated village construction (both of materials used and layout of the village), and the sharing of water resources. Without doubt the rapid growth of Leh and now the region, especially in the last 15 years, is changing this historical way of living at the high elevation of the region, bringing with it plastic pollution, air pollution, noise pollution, and other forms of trash that are part and parcel of consumer-based petrocultures.

Furthermore, the shift from an agrarian to a tourist-based economy may make it harder for the region to generate resiliency in the face of climate change. This is exacerbated by the same trend lines lamented by the US agrarian writer Wendell Berry in his 1977 treatise *The Unsettling of America*: with the mechanization of agriculture, youth are out of jobs and out migrate to the cities where there are more job opportunities and things are flashier, faster, and seemingly more entertaining. This process is repeating itself in Ladakh, with many youth leaving traditional farming hamlets and out-migrating to Leh, or down to Delhi where they attend university and often remain. Many of the village elders with whom we communicated during our class visits lamented this loss of generational traditional ecological agrarian knowledge, especially in the face of climate change.[4] The result is that today there are fewer people engaged in what had been largely self-sufficient farming—barley, apricots, cows [dung and milk, both], yaks, peas, and others—and instead more Ladakhians are becoming increasingly dependent upon imported food and clothing items that are supported by governmental subsidies. Meanwhile more people are shifting toward catering to construction, long haul trucking, tourism, and trekking/river rafting (industries that are utterly dependent upon the use of fossil fuels).[5]

There are other layers in providing context for embodied religious resiliency in Ladakh. Besides recognizing that changes in technology are never socially, politically, economically, or ecologically neutral, these shifts are taking place within a larger Indian/South Asian context of neocolonialism (first British, and now transnational corporation) that is reshaping not only the landscape but also the social fabric (Gadgil and Guha 1995). The people impacted by these shifts in land and culture-scapes are grouped into three separate categories by the ecologist Madhav Gadgil and historian and economist Ramachandra Guha. The first category are omnivores—those following the pathway of petroculture, embracing a lifestyle based on fossil fuels and consumption of the earth's resources. In Ladakh these are visiting Indian tourists, inspired by *3 Idiots*; ex-Israeli servicemembers enjoying "Shangri-La"; Global North trekkers/climbers/river runners/tourists visiting monasteries; and a rapidly emerging owner/entrepreneur class capitalizing on catering to these three audiences. The second category they label ecosystem people: fisherfolk, shifting cultivators, and others who meet most of their needs via the local environment and regimes of traditional ecological knowledge. In Ladakh these would be those engaged in a traditional Buddhist/agrarian economy of bio-ecological place. The third category are the displaced victims of enclosures precipitated to keep omnivores fed—the category of urban poor they call ecological refugees (1995, 2–4). Such refugees (Hindu, Buddhist, and Muslim alike migrating from various parts of India) are increasingly appearing in Leh where they undertake the labor of building new hotels and guest houses. Yet as they point out, "As the natural world recedes, so shrink the capacities of local ecosystems to support these people" (1995, 3–4). This capacity is now impacted by the threat multiplier of global warming, where at high elevations subtle shifts of temperature and precipitation patterns can have amplifying impacts on human lifeways.

This is especially so for the biocultural lifeways of ecosystem peoples—those best positioned to anchor resilient, adaptive responses to regional climate changes. For example in India, in sedentary, nomadic, and semi-nomadic biocultural caste groups organized around the unit of a village, made of several endogamous caste groups, "caste society had developed an elaborate system of the diversified use of living resources that greatly reduced inter-caste competition, and very often ensured that a single caste group [through hereditary rights] had a monopoly over the use of any specific resource from a given locale" (Gadgil and Guha 1992, 95). This form of biocultural organization emerged in the Indian subcontinent during the Gupta and post-Gupta period (300–800 CE) where local elements, seeds, animals, plants, forests, and rivers became sacred in folk religions. What is key is that "This belief system had a clear role in regulating and moderating the use of natural resources. It legitimized in a new framework the protection accorded to certain elements of the landscape . . . as sacred to a variety of deities . . . . Even more important than religious sanctions, in this regard, were social conventions [that] helped resource conservation" (Gadgil and Guha 1992, 104). Similar dynamics of social conventions aiding resource conservation by traditional ecological knowledge-and-dwelling biocultural systems are seen in ecosystem peoples the world over: from water temples and village managed bio-ecosystem rice water practices in Bali (Lansing 2007), to West African rice cultivation (Richards 1996), to Carolinean fishing in Micronesia (Akimichi 1996), to strategic forest farming and forest cultivation for biodiversity in the Amazon (Moran 1996), to the high-elevation snow-fed agrarian farming villages of Ladakh. Climate change threatens all of these bioculturally evolved systems via forced out-migration (climate refugees), changing seasons and weather patterns (and thus farming practices in response), changing hydrology and water availability, and changing migratory patterns of other species.

It is important to not misrepresent ecosystem people in Ladakh as ecotopians who were living a perfect, prelapsarian life until Western contact and the tourist gaze. Even if there was environmental protection in the region (and there was), we need to recognize: (1) "whilst non-industrial peoples may have lived in ways which were more environmentally sustainable than contemporary Western [or any form of omnivore living, anywhere] lifestyles it cannot be assumed that this was a result of their holding environmental values"; and (2) 'whilst many religio-cultural traditions do have a strong connection with the natural world, elements of the natural world may be considered as sacred without any explicit consciousness about the relevance of this to an environmental crisis" (Tomalin 2002, 15).[6] Rather, the Tibetan and tantric Buddhism of Ladakh is built upon recognizing chthonic demons, spirits, ghosts, and other agental beings of place and spiritual powers that can impact the individual, communal, and natural health of a town/village and its surrounding valley. Important dramaturgical rituals of protection are performed by Buddhist monks, with these rituals occurring in homes of laity, in monasteries, and at sacred sites. Lay people also maintain their own domestic shrines and perform daily dramaturgical rituals of protection, often based on repeating *mantras* and offering incense, food, and libations to household Buddhas, ancestors, and other powerful deities. Towns, monasteries, valleys, and geographic regions are also protected by *tulkus*, "the human emanations of either celestial bodhisattvas or transcendent reincarnations of Buddhist teachers" (Butcher 2013a, 17).

In many ways, then, the Ladakhian Buddhist cosmology is one of Buddhist gods, powerful Buddhas, and enlightened or powerful monks acting to protect villages and valleys from demons and less powerful spirits. The landscape of Ladakh is dotted with *stupas* that help contain and control these dangerous spirits, as well. When seen in concert with the monasteries asserting their celestial and earthly power by their location high atop mountains, where there's power and purity, it is a choreographed landscape of emplaced biocultural meaning. The biocultural meaning for Ladakhians, at least prior to the onset of modernity, is that Buddhist monks and Buddhism play a central role in protection, in embodying and teaching the *dharma*, and in regulating *karma* in the community. There are other forces at play which must be pacified, and these forces can cause floods, mudslides, droughts, and other natural disasters. Thus the role of the various Buddhist monastic lineages in protecting people and place by maintaining ritual and moral purity is central to the biocultural lifeways of Ladakh.

At least, it was. Now there is tension because Western discourses of science brought by tourists, or local elite trained in universities in Delhi or abroad, or by governmental development agencies, are competing with and at times displacing more traditional Buddhist cosmological understandings of power and place. If it is rising greenhouse gases causing changing snow and rain and cropping patterns, and not spirits that need to be controlled and pacified by Buddhist monks, then what happens to the prestige and influence of the monasteries? What happens to the efficacy of the *dharma*, and the need to be concerned with moral purity to abate *karmic* retribution by spirits, if geology, physics, chemistry, and biology offer causal explanations? Currently the efficacy of monastic ritual and action in pacifying spirits and forces, and thus in keeping ritual and moral purity, still has explanatory power (Butcher 2013b). This is backed up by conversations my class had with locals on changing weather patterns—elders specifically saw this through the

Figure 6.1   Farming Valley of Alchi, Ladakh. *Source*: Photo by Author

Figure 6.2   Water Being Diverted to Various Fields, Alchi, Ladakh. *Source*: Photo by Author

lens of *karma*, where the young leaving the tradition and the moral impurity brought by tourism were angering the spirits of place with retribution coming through changing weather. However, those we encountered who have been educated at universities, or who represented various environmental NGOs, or who accessed international websites that covered climate change, very clearly attributed changing weather patterns to global warming. In total, all the people with whom we interacted expressed worry—no one denied the weather is changing, and all noticed the impact of changing patterns (timing, amounts) on snow and rain, agriculture, and seasonal patterns of various flora and fauna.

This is borne out by a study of six villages in Tibet where Anja Byg and Jan Salick found that all villagers noticed changes in weather due to climate change (2009). They attributed this to either material reasons (more driving, cutting trees) or spiritual reasons (impure moral actions, not following rituals, angering deities by blowing up mountains and treading on sacred places, behaviors of outgroups [tourists, other villages], increased plastic pollution).[7] However, what is key for this study is that the participants had never heard of climate change despite the Chinese government having climate policies; and while perceiving changes in weather, they understood this as a local phenomena caused by the above reasons. Even without this scientific understanding,

Figure 6.3   Signs at Takmachik, Ladakh. *Source*: Photo by Author.

though, "Tibetan villagers are concerned about the same issues as scientists . . . such shared concerns may constitute common ground where scientists, policy makers and villagers' interests converge. At the same time it is, however, also necessary that villagers' concerns about the spiritual implications of climate change are taken seriously even where these are not shared by the scientists and policy makers" (2009, 165). Similar phenomena are occurring in the rural parts of Ladakh, at least with elders in villages—yes, things are changing, and elders largely attribute these changes to similar material and spiritual causations. Yet, as the weather continues to shift, and impacts become more pronounced and maladaptive, education about the science of climate change, especially through Buddhist monks, will most likely increase. Indeed, some Buddhist leaders in Ladakh are beginning to include climate change in their *dharma* talks, ritual programming, and community outreach and are slowly mobilizing to prepare for the predicted impacts of climate change.

This adds another layer in this chapter to its focus on embodiment, and thus bodies and religion. Here adaptive responses (described below) are borne in

Figure 6.4   Remains of Ice Stupa, Takmachik. *Source*: Photo by Author

part because of lived, bodily experienced changes in agrarian-based villages. Ladakhis physically with their bodies see, taste, feel, and hear changes in the environment. They bodily experience these changes of different weather patterns, animal migrations, insect outbreaks, and warmer temperatures and less snowfall. They are also bodily concerned—for many, especially elder farmers and pastoralists, these changing weather patterns threaten their ability to grow food and have enough water. In short, their bodily quotidian needs are already unmet in some seasons, and this is of concern. That most causal explanations are still grounded in Buddhist cosmologies suggests that a regional Buddhist framing of climate changes will be needed to help both ground and communicate and then biophysically generate climate resiliency. And this is indeed the case, for example, in October 2019 with Chetsang Rinpoche, the 37th throne holder and head of the Drikung Kagyu order of Tibetan Buddhism, joined with the local BJP MP and local NGOS to advocate for sustainable development in the region that can adapt to climate change. Part of this advocacy

Figure 6.5   Reservoir in Takmachik where ice melt from the ice stupa (to the right of the reservoir tank) is stored to be used by the village to water spring plantings. *Source*: Photo by Author

includes a call by regional Buddhist leaders to plant a billion trees in the Himalayas, including in Ladakh.[8] The initiative is called "Green Himalayas" and is based at a 250-acre site in Phobrang, Ladakh that will serve as a model of sustainable development, with partnerships between NGOs, leading businesses, the government, and Buddhists, with a focus on empowering women. This site is intended to serve as a sustainable development seedbed and model that other communities throughout the Himalayas can emulate. Of note is that Rinpoche Chetsang is also the founder of one of the NGOs leading the project, Go Green, Go Organic, which focuses on organic farming practices in Ladakh.

Rinpoche Chetsang's religious (and thus, moral) standing in the region afforded by specific factors related to his status as a reincarnated lama/teacher of his lineage affords a level of prestige and charisma that can, and increasingly does, motivate lay followers, businesses, and governmental leaders in Ladakh to proactively respond to climate change.[9] He is also empowering monks within his lineage to become engaged in climate adaptation and resiliency. This includes organizing tree planting, supporting organic farming, constructing ice stupas, and creating miniature ice glaciers. It is to the last two that I turn to conclude this section on religious resiliency, as on my trip I witnessed the remains of both an artificially constructed (by bodies!) ice stupa

Figure 6.6 *Stupas* on the Landscape, Ladakh. *Source*: Photo by Author

and ice glacier, respectively; and also because these are adaptive responses to already felt water shortages triggered by climate change.

The construction of ice stupas and creating miniature ice glaciers are two strategies originally formulated by local engineers to hold water longer for villages and seasonal watering of crops. If climate change means less snow and earlier melt, that means water is not as readily available for spring and early summer planting. This means the farming practices, including the seed varietals, developed as traditional ecological knowledge and that depend on water at certain times and in certain amounts are not able to get such water. The lack of water then threatens food supplies and livelihoods. By freezing water so that it is released slower, and in some villages, into catchment ponds, the water is temporarily available for agricultural and other village needs closer to the time it would have come before the onset of global warming. Ice stupas and glaciers are therefore engineered bio-adaptive technological strategies aiming to help generate climate-adaptive interventions in response to a changing hydrology (Clouse 2016).

Figure 6.7 Himalayas in India, Covered in Snow in June, 2019. Predicted impacts of climate change suggest much less snow in the years to come will cover these mountains, impacting various pilgrimage sites and the livelihoods for hundreds of millions. *Source*: Photo by Author

One idea for an ice stupa emerged from Ladakhi engineer and social visionary Sonam Wangchuk, with the first one built in the winter of 2013–2014. Around the same time the civil engineer Chewang Norphel, aka "Ice Man," also had the idea and created his own.[10] The engineered water management strategy marries cultural prestige (Buddhist iconography and import of *stupas*, which are central to Buddhist practice) with physics, so that "The ice stupas of Ladakh represent a new type of climate-adaptive landscape in the region" (Clouse 2017, 248). However, the location of these hydrosocial artifacts, even when built with the blessing of Rinpoche Chetsang, are not without contestation, especially when they impact customary access to streams and water (Sharma 2019). As with embodiment everywhere in material biocultural places, issues of power and access are always present, whether at a surf break or in the Amazon (see next chapter), in North Dakota, or in rural Ladakh!

Longitudinal studies are needed to see both how much water resilience these artificial glaciers and stupas can generate; but also how the science behind them (the climate science showing the need for new water regimes, and the science of building them) may be grafted onto existing religious life-ways. On the latter, three local artists in their 20s had Sonam Wangchuk's blessing to carve into the engineered ice stupa an actual *chorten*, or *stupa*.[11] These artists recreated on the inside of the artificial stupa a model of *Changchup Chorten*, a *stupa* that symbolizes the enlightenment of Buddha. The artists claim they undertook the art piece to help locals bridge the divide between the science of the ice stupa, with the iconographic power of a traditional Buddhist *stupa*, so there would be a cultural container within which to appreciate the larger ice stupa.[12] But this grafting is strongly suggestive of the future of religious production in the Anthropocene/Carboncene. Religious bodies will be trying to generate biocultural resilience in the above and other ways the world over, more and more, and as scholars we need to attend to this with our research regimes. Buddhist monks holding rituals around ice stupas that mirror and function the same as those held at other sacred Buddhist sites, including of other *stupas*, is a new religious phenomenon. The dramaturgical bodily action of pilgrimage to these sites, of placing prayer flags and burning incense and performing dance and ringing bells and chanting *mantras*, of lay people bringing their bodies and the bodies of their children to participate in these rituals—this is the fertile ground of new religious production and the performance of new ways of embodying religion in an era of climate change.

That the bodily actions here are part of adaptive responses to changing weather patterns that threaten livelihoods needs study. This becomes even more evident when those studied by our field themselves express the religious import of this biocultural bodily adaptation to climate change:

> The monastics spearheading the artificial icefall glaciers do not view organizing these projects as separate from their religious practice so much as an extension of it. As one monastic leader describes, while traditionally monastic practitioners would spend their time performing rituals or engaged in meditation, with the myriad issues facing the world today, it is necessary for monks and Buddhist practitioners to do more than meditate or recite mantras. In his words, "Today, we are monks in action!" . . . He [i.e., Rinpoche Chetsang] says that religious leaders like himself can no longer just stay in the monastery and pray for a better world, but rather must also "go out into society and help people according to the needs of the twenty-first century." Although Chetsang Rinpoche's monastic and lay students in the Indian alpine desert region of Ladakh are neither numerous nor affluent, a large number have followed his call and become very active in local efforts to protect their environment and adapt to the localized impacts of the climate crisis. (Yonnetti 2020)

Such adaptive actions are by definition embodied: physically planting trees, the ritualized creation of ice stupas and glaciers, and the building of new water pipes from ice stupas and glaciers to fields. The material agency and reality of our planetary home is changing, and humans-in-ecosystems via the vehicle of religion are adapting. As seen in the arid elevations of Ladakh, the adaptation is being driven in part by Buddhist monks and their engagement with their communities; by concerned local engineers and farmers; and by a fusing of traditional ecological knowledge with Western science. When the efforts at biocultural adaptation to climate change are dramaturgically performed in religious contexts, then these religious facts become the required future of religious studies to be studied, analyzed, and then taught in classes.

So far, examples drawn from Hinduism, Islam, and Buddhism remain within conventional boundaries of world religions. However, religious studies need not only attend to such evidence. The onset of nascent, post-materialist adaptive responses to climate change also presents a rich area of exploration. I turn now to a final chapter that contains three case studies that provide examples of emergent posthuman performances of religiously influenced resiliency in post-materialist Global North enclaves.

NOTES

1. https://www.stockholmresilience.org/research/planetary-boundaries.html Accessed May 18, 2020.

2. For a deft analysis of the impacts of increased tourism on the ecology of the region, although that somewhat perpetuates romanticized notions of Leh and Ladakh as a Shangri-La that was always sustainable, see https://www.buzzfeed.com/ariel bardi/what-has-3-idiots-done-to-ladakh. Accessed May 25, 2020.

3. For a much more nuanced and contemporary exploration of the biocultural impacts of climate change on the region written by a Westerner, see Jonathan Mingle, *Fire and Ice: Soot, Solidarity, and Survival on the Roof of the World* (2015). I thank Zeff Bjerken for gifting me my own copy of this enlightening book—one I highly recommend the reader if they want to see the nuances and tensions unfolding in the region as presented via an astute mix of ethnography, science, and autobiography.

4. A lament largely echoed in the ethnographic stories shared by Mingle in his book.

5. For two documentaries that expertly explore these changing cultural, technological, climatic, and emotional terrains, see Stanzin Dorjai Gya's *The Shepherdess of the Glaciers* (2016) and *Jungwa: The Broken Balance* (2012). To access both visit https://www.stanzindorjaigya.com/.

6. We must also be leery about the move from the Dalai Lama to frame Tibetan Buddhism as "green," as this move from the 1980s and 1990s is less a reflection of intrinsic environmental concern (itself a Western outgrowth), but is a move by a cadre

of elite Tibetans in exile to help access Western resources, especially of the financial kind (Huber 1997). While the Dalia Lama does not speak for all types of Buddhism in Ladakh, the larger gestalt of seeing Buddhism in this part of the world as green is impacted by such narratives disseminating directly from high ranking monks and their marketing teams in Dharamsala.

7. Similar dynamics were found in Hindu Kullu Valley in the Western Himalayas by scholar of Hinduism Ehud Halperin, where villagers blamed immorality and sins like abandoning cows, as well as population growth and increased vehicular traffic, for observed climate changes (2017).

8. https://timesofindia.indiatimes.com/india/project-launched-to-make-ladakh-r esilient-to-climate-change/articleshow/71468275.cms Accessed May 26, 2020.

9. Here's one brief biography of him from a blog celebrating his leadership efforts: "In many of the villages in the Changthang region of Ladakh that surrounds Pangong Lake, environmental projects are being undertaken largely due to the inspiration and leadership of the Tibetan Buddhist teacher Kyabgön Chetsang Rinpoche (b. 1946). Honored as a United Nations Mountain Partnership Ambassador in 2015 for his efforts to promote sustainable livelihoods and environmental stewardship in High Asia, Chetsang Rinpoche is the leading Tibetan Buddhist teacher in much of this area of Ladakh which follows the Drikung Kagyu sub-school of Tibetan Buddhism. During the course of his public teachings outside of as well as in the region, Chetsang Rinpoche has often stressed the need for people to each take whatever small steps they can in order to help combat the intensity of climate change globally as well as to be proactive in adapting to its anticipated effects in their local settings." From http://blog.usief.org in/Posts.aspx?PostID=2145 accessed May 26, 2020. It is unclear how much his views of climate change are informed by scientific discourses and understandings. Further ethnographic research into this question by scholars, ideally via interview with Rinpoche Chetsand directly, would help tease out these complexities in regard to his proactive commitments to sustainability and to stopping climate change.

10. https://www buddhistdoor.net/news/ice-stupas-address-springtime-water-shortage-in-indias-far-north Accessed May 26, 2020.

11. The ice stupa is called such as it resembles the formation of a *stupa*, with a wide base and smaller top, like a pyramid. Calling it such adds cultural legitimacy. However, ice stupas are not such unless blessed by monks. The three artists carved out the base of an ice stupa and inside it carved what would be seen by viewers an "authentic" *stupa*, although one not blessed by monks, or made by monks.

12. https://www.buddhistdoor.net/news/ice-chorten-in-ladakh-becomes-a-budd hist-symbol-of-the-climate-change-crisis Accessed May 26, 2020.

Chapter 7

Post-Materialist Posthuman Dramaturgies and Resilience

This[1] chapter is in part anchored by a study undertaken by me, religion and nature scholar Bron Taylor, and psychologist Jen Wright (2020). The data we generated strongly suggests that more sustainable responses to deleterious environmental behaviors can occur and in the near-term will more readily occur by people holding certain types of religious identities. Our research found, to levels of statistical significance, that those who have environmental and religious humility, accept evolution, hold a holistic worldview, and who hold varieties of dark green nature-as-sacred religious identities and views are much more likely to report that they engage in more sustainable behaviors compared to those who are monotheistic and anthropocentric in their religious and ethical bearing.[2] To generate this finding we developed a globally unique survey scale that combined questions about environmental humility, religious humility, dark green religious perceptions, and views of evolution and triangulated these with proenvironmental behaviors. These are self-reported choices related to food, energy, and transport and that utilize renewables, public transit, and vegetarian diets—behavioral choices lower on the "carbon food chain," and thus more sustainable.[3] This research provided further evidence that "the most dramatic environmental mobilization is not emerging from the world's predominant religions but in a host of ways among those who are converging toward worldviews grounded in evolutionary and ecological understandings" (Taylor 2016b, 224). Given this, the case studies in this chapter focus on emerging post-materialist and posthuman identities and practices in the Global North. These identities require more sustained study by the academy and in religious studies, as well, as they point toward real-time shifts in human concepts of self and interactions with bioecosystems and ecologies of place in the context of runaway climate change.

My focus is grounded as well by the passage shared in chapter 4 by Amitav Ghosh which I said I would revisit later: "one of the uncanniest effects of climate change [is a] renewed awareness of the elements of agency and consciousness that humans share with many other beings, and even perhaps the planet itself" (Ghosh 2016, 84). This insight is related to findings on dark green humility (above), to religious dramaturgical bodily performance, and to concerns for resilience and sustainability. As more humans recognize the fragility and brittleness of our lifeways, and come to realize we are utterly dependent upon the rest of life and its flourishing independent of and also dependent upon our human agency, then Ghosh's insight holds: some humans will likely come to realize the agency and consciousness of nonhuman others and how they co-create our sense of self and well-being with us, individually and collectively. Of course, some human biocultures have never lost this awareness, as seen with indigenous leaders leading #NODAPL protests against petroculture.

Those highlighted in this chapter are grappling with what it means to share life with an animate earth (Harding 2006) in ways that are culturally, geographically, and religiously specific. These examples provide evidence of emergent posthuman ways of enacting religion via bodies in places, and especially in post-materialist (i.e., economically wealthy with easy access to abundant sources of food, clothing, and shelter to meet basic needs) spaces. The case studies provide evidence as well for the observation from gender theorist Stephanie Clare who shares that "The lived body as point of view is located, but not fixed. It is not an object but a responsive gearing to the world [where s]ensations are entangled with how we move and act in (and with) the world" (2019, xxvii). How are emerging forms of posthuman bodily expressions where the natural world is understood as the location of the sacred moving in and acting with the world, driven by environmental and sustainability concerns, in an era of climate changes? What are these bodies doing, that may be religious, and why?

The analyses of these "anticipatory communities" (Rasmussen 2013) must be cursory, pulling largely from web-based archives (YouTube videos, websites, posted interviews, magazine articles, documentaries). However, the point here is not to undertake rigorous fieldwork on religious materialist moves toward embodied resilience in climate changes—my hope is that the field, and the academy at large, undertakes this needed work instead, leading to insightful and engaging monographs that can better point all of us toward examples of religiously performed regenerative and resilient thriving in ecologies of place. Rather in this chapter I am quickly pointing toward potential "dramaturgical bodily practices that are religious" as these emerge in the face of climate changes, providing snapshots of what religious "bodily hermeneutics of climate changes"[4] may resemble. In these vignettes I point toward those humans and their bodily practices who act in creaturely solidarity with the symbiotic real, moving beyond the severing of "the Human" from the rest of life (Morton 2017); I point toward those who are instead "rooted and rising" (Schade and Bullitt-Jonas 2019) in a posthuman religious, bodily engagement with the planet.

Given the above, I do not view the examples contained in these brief case studies as forms of ecopiety, where the virtue signaling of "green consumerism" (what religion scholar Sarah McFarland Taylor calls "consumopiety") by post-materialist liberals is a hallmark of a religious environmentalism (McFarland Taylor 2019, 3–4). Rather, for me these examples are part of the larger, global and slowly maturing vanguard of biocultural lifeways within post-materialist industrial societies—or peasant and indigenous societies forced to deal with the unasked for violence of climate changes wrought by industrial societies, as seen in the last chapter—that are generating resilient practices focused on regenerating life. These are signals of emerging transformative physical actions of those living within the hypocritical absurdity of late fossil fuel-based capitalism, who recognize they must change how they live, and do so from within the compromised Capitalocene/Carboncene. To me they present evidence of a preliminary cultivation of Timothy Morton's ecological thought, referenced in chapter 3: an ecological awareness that inspires coexistence with nonhumans, in the types of symbiotic and sympoietic relationships theorized by Donna Haraway.

The future of religious bodily action 20–30 years out will most likely resemble many of these same types of "new ecological civics" (Northcott 2015, 43) that move beyond nature/culture dualisms. These bodily actions that fuse ecological civics with nature-as-sacred religiosities/spiritualities, then, are part of the emerging move from what is to what if (Hopkins 2019) in a transition to life after oil (Boudinot and LeVasseur 2016). Most of these efforts at transformation, and for sure these three from affluent Global North contexts, will occur from within dominant petroculture, omnivore lifeways. As such, those attempting to generate new lifeways from within such White CisHetero Settler Colonial post-material contexts will carry with their efforts privileges and power dynamics that are unavoidable,[5] and that will impact and shape their transformative religio-ethics and dramaturgical practices. There will also always be levels of layered hypocrisy, both materially and epistemologically in these efforts at generating resilience. This is because those in post-materialist contexts have the privilege and ease from which to critique empire-based petrocultures and dominant religious lifeways that are perceived to be at odds with a queered, full-spectrum, posthuman flourishing. To use Gadgil and Guha's terminology those who are empty bellied or who are ecological refugees do not have such luxury—their basic material needs of food, clothing, and shelter are not met, or if met are constantly stressed and not dependable. Rather, in the turn toward exploring queer posthuman materialist dramaturgical religious resiliency in post-materialist Global North enclaves undertaken in this chapter, the groups I explore below would be Gadgil and Guha's omnivores. Their nature-as-sacred religiosities will therefore be influenced by such petrocultural metropole positioning.

SUSTAINABLE STOKE

Evidence suggests that prosustainable efforts are developing in various types of outdoor recreation practices. This is because those engaged in deep contact with ecologies of place and patterns of bio-ecological unfolding in these dwelling places often report phenomenological religious experiences triggered by the more-than-human world. These embodied nature-as-sacred experiences for some lead to efforts to preserve such bio-ecologies and to generate more prosustainable lifeways. So while here I explore such emergent dramaturgy in surfing, the reader should keep in mind this is a focusing mechanism for larger patterns emerging in other nature-as-sacred practices for those also engaged in fishing, hunting, kayaking, rock climbing, windsurfing, hiking, camping, and similar outdoors activities.

The title for this section comes from a fantastic book of the same name, edited by Gregory Borne and Jess Ponting (2015), two leaders in the emerging field of critical surf studies (Hough-Snee and Eastman 2017). In it they document how surfers around the world are undertaking, and at times, succeeding, in a variety of sustainability campaigns. These range from cleaning the ocean from sewage and plastic, to protesting long line fishing, to helping create new models of and markets for cradle-to-cradle surf fashion, wetsuits, and surf boards (see also Borne and Ponting 2017). This is no small feat, as despite stereotypes of surfers as being anti-establishment "soul rebels" who just want to hang out at the beach, the reality is that surfing is big business: billions in sales annually on items ranging from flip flops to bathing suits ("baggies" in surf parlance) to surfboards to wetsuits to GoPro cameras that are used to film surfing from the water. Add to this the (pre-COVID) huge carbon footprint of surfers traveling the world in search of waves (similar to the pilgrimages described in chapter 5) and professional surfers traveling the world for contests and the reality is that surfing as both lifestyle and industry is not ecologically benign. There is also now a concerted effort to target urban dwellers living near coastal areas the world over with surf lessons, where those in their 40s and 50s with expendable income and young children are marketed a lifestyle of health and ocean engagement. The economics and eco-footprint of this emerging demographic will only add to the larger impact of surfing on planetary ecosystems and climate health. Any assessment of bodily practices in service of resilience at the interface of surfing and spirituality qua religion must take this context of modern surfing enabled by petroculture into account.

Surfing itself is part of a larger water knowledge complex system, akin to the dwelling perspective described by Tim Ingold. There is a long history, going back centuries, of water borne, embodied, ritualized, place-based knowledge that came with mastery of indigenous watercraft technologies, especially in the African and African diasporic worlds (Dawson 2018), as well as Polynesia, and also in the Caribbean, among other bio-ecosystem places. This was at a time when Europeans, largely, did not enter the water for leisure activities. This same embodied traditional ecological knowledge of the ocean and its waves was central to Hawaiian culture, and thus religion, as the two were inseparable. It was not until colonization of Hawai'i by the United States and the strategic dismantling of the Hawaiian social and royal system, which included missionaries prohibiting surfing, that white people were able to claim surfing as theirs. However, surfing was never not practiced by Hawaiians (Moser 2016; Walker 2015; Clark 2011), and the growth of surfing and its image as a white, anti-establishment, rebel, beach-based culture (Lawler 2010; Flynn 1987) belies a long and ongoing history of colonization and dispossession (Laderman 2014) and rampant heteromasculinity (Evers 2019; Waitt 2008; Comer 2010).

That said, of the 17–35 million global surfers (numbers estimated by the Surf Industry Manufacturer's Association) there are many who claim surfing as a spiritual practice. I in fact label myself as one, as I "converted" to the Church of Open Sky and Ocean back in the early 1990s. Bron Taylor ably describes surfing as a form of aquatic nature religion (2007) and in a passage that parallels some of the key arguments being made in this chapter argues

> that a significant part of the evolving global, surfing world can be understood as a new religious movement in which sensual experiences constitute its sacred center. These experiences, and the subcultures in which people reflect upon them, foster understandings of nature as powerful, transformative, healing, and even sacred. Such perceptions, in turn, often lead to environmental ethics and action in which Mother Nature, and especially its manifestation as Mother Ocean, is considered to be worthy of reverent care. This produces a holistic axiology that environmental ethicists label biocentrism or ecocentrism. Surfers' deep feelings of communion and kinship with the non-human animals they encounter during their practice, which sometimes take on an animistic ethos, can also lead surfers to discrete political action on behalf of particular species and individual animals. (2007, 925)

Here Taylor underscores what many surfers will explain and witness to, if pressed, about why they surf: the embodied feeling of being in the ocean, riding along liquid waves of energy, as a form of ecstatic religious practice. This is not to say that other motivations are not present for surfers: competition, posturing, trying to de-stress after a hard day of work, getting exercise, being with friends, trying out a new board, and many others. Yet this assumes there is always a pure, singular motive for any religious practice, which is a simplistic understanding of why people engage in and perform diverse religious activities.

There's a tension, too, that must be highlighted, whether the bodily dramaturgy is surfing as a form of nature-as-sacred practice, or hiking, or kayaking, or mountain climbing, or a variety of other nature-based bodily practices. This is that between individual experience, often afforded by petrocultures, and larger systemic behavioral effects. To date surfing and other recreational nature-based practices (which for some are spiritual/religious, as focused upon in this section) often privileges the former, whereas the latter, at least in support of prosustainable behaviors, is largely lacking. And for those invested more on the individual side, where they may have peak experiences that are bodily spiritual and/or ecstatic, this often will not translate into a posthuman, queering of human identities. So this foray into surfing spirituality is aided by extant literature to support my analysis, but these tensions between individualism (especially in post-materialist, omnivore settings) and climate adaptation at larger scale and within the context of a posthuman hermeneutic, especially over time, need to be teased out by further studies.

Besides an experiential element, surfing also has its own set of religious archetypes: an *axis mundi* (the "mecca" of "the Seven Mile Miracle" on Oahu's North Shore); ritualized paddle outs for funerals; rites of passage where grommets (young surfers) receive hazing; pilgrimage to sacred surfing sites and going on "surfari"; a pantheon of surfing heroes and heroines; a sacred origin story (in Hawai'i and seeing it as the sport of kings); a robust and varied material culture; and its own sacred texts (various journals and documentaries, including the famous surf movie *The Endless Summer*). Those who interact with and dwell in surfing subcultures will, over time, be exposed through socialization to all of the above.

Those who surf, who spend countless hours in the ocean, may come to have self-described spiritual experiences. This provides the touchstone and entry into my lens of religious bodily dramaturgy. These surfing-based performative bodily experiences are a mix of ritualized practice (getting gear into the car and driving to the beach, putting surf wax on the board, putting on a wetsuit, putting on a leash, paddling through the lineup out past the breaking waves, stroking into and catching a wave and then getting to one's feet, riding down the line of a breaking wave, kicking out, paddling out to do it again, putting the board back in the car after the session, leaving the beach) and hard-won synesthetic satori-like trance states, where muscles align with stilled brain and reading of shifting swells of water and tide and wind that culminate in balanced 5–20 second rides of bliss, with the body surrounded by oceanic colors, sounds, and sensations. Surf speak is ripe with religious terminology, all tethered to the performative, bodily actions and practice of the act of surfing itself.

But however enjoyable, the embodied and performed practice of surfing is not detached from the realities of climate change. A changing climate will have multiple impacts on the oceans and seas, the ecologies of place where occur the biocultural religious practices of surfing. For one, ocean acidification will impact oceanic food chains, altering the range and diversity of ocean organisms. Data suggests there will be much less biodiversity in the ocean, meaning surfers will encounter and interact with an impoverished level of ocean wildlife. For many surfers, seeing and encountering and even interacting with dolphins or stingrays or sea turtles or fish or manatees or whales (or sharks!) are a key part of this religious practice, and these ecological more-than-human encounters that many consider a key sacred bodily experience that is part of surfing will most likely be reduced. The physical waves of liquid energy that are required for surfing will also be impacted by climate changes. Waves are caused mostly by wind, and many surf spots depend on seasonal weather patterns that produce certain types of waves that come from certain directions and thus break in certain ways. This dependability adds to the formation of ocean-dwelling knowledge in surfers. For example, the North Shore of Oahu in Hawai'i receives its famous waves always in the winter, the result of major storms in Alaska that send 10–20 foot waves from certain swell angles toward the reefs of Pipeline, Backdoor, Off the Wall, Rockies, and Waimea Bay.

My own home break does best on hurricane swells (see Figure 7.1), and then winter northeasters that blow across the Eastern seaboard of the United States. Climate change will cause more extreme hurricanes, bringing a level of increased threat to my break, and washing away sand so that the waves break differently. And from when I began surfing in the 1990s, I have personally noticed a difference in winter waves on the US East Coast—we no longer get them as often as we used to, as the northeasters we used to get do not happen much anymore as weather patterns near the Great Lakes shift. Ocean currents are shifting, too, which will impact wave production. And of course there's sea-level rise—many breaks that currently exist will simply cease to exist at some point in the future. What then for the religious practice of surfing? To dramaturgically practice this nature-as-sacred practice with its attendant regional subcultures requires a beach, a headland, a reef—some ecology of place for actual, physical waves to break so bodies can dance and glide upon them in religious supplication. And then places to congregate to discuss surfing and its import in people's lifeways. Some say

Figure 7.1   Author, Surfing Hurricane Isaac Swell, August 2012, Folly Beach, South Carolina. *Source*: Photo by Ben Roth

that wave pools and wave tanks, where machines artificially create waves in a human-created environment, may be a replacement, but for many surfers this leads to an act of wave riding but not surfing;[6] while similar in feeling, it is a "fallen" simulacra based religious practice, not the same as surfing in the ocean, where our animal bodies are immersed in the more-than-human liquid world.

Given the above, what can an analysis of surfing bring to insights into the body and religion, and dark green humility; to affective responses to nature and embodied care for the earth in an era of climate change; to understandings of evolution, environmental ethics, and concern for the climate future? Insights from religious studies professor Robert Fuller (2013) and his unpacking of religion in America in the twenty-first century are useful here. The following description captures well the types of embodied religion that surfing represents:

About 18 percent of Americans consider themselves personally spiritual even though they don't have a formal religious affiliation. This group has little confidence in organized religion [but many] pursue religious interests and for this reason are often characterized as exhibiting "seeker spirituality" or "quest spirituality." . . . they self-consciously explore as ever-changing interests dictate. Those who consider themselves spiritual, but not religious, are also the most likely to expect religion to have a pronounced experiential dimension. (169)

Here we can place surfing as a type of quest spirituality,[7] and at its core is a pronounced experiential dimension: of the ocean, of gliding, of being immersed in liquid, of paddling and standing and submersion, of sunsets and moonrises and sunrises, of scented wax being rubbed on a board, and dolphins swimming by and coral reefs being under feet and shells on the beach and sand between the toes.

For example, the Australian multiple-times women's world surfing champion Stephanie Gilmore talks about surfing as a form of escape, where "the soul of surfing for me is about having a strong connection with the ocean; it's like finding a soul-mate when you go out and surf. Getting barreled [being immersed in the curling part of a wave] is a full connection right there; you're able to bond with a part of the wave and it's the best thing in the world. All the women who have influenced me, they seem to have a strong soulful connection to the ocean in a way" (McCoy 2014). Meanwhile, "Mr. Pipeline" Gerry Lopez, famous for bringing 1970s New Age yoga calm to the death waves of the Bonzai Pipeline (surfers literally die there), states that "I believe that surfing is a spiritual dream. It's extremely peaceful. Nothing else matters but that place in time. Everything is non-existent but that moment" (McCoy 2014). The narrators of this same documentary by famed Australian surf filmmaker Jack McCoy conclude that "Aloha and deep respect for the ocean are the heart and soul of surf culture" (2014). Other surfing world champions feel the same. The deceased Andy Irons from Hawai'i was a three times men's world champion who described exiting the barrel with an explosion of the wave as "being kissed by god," and 11 times men's champion Kelly Slater from Florida shares that "Surfing is my religion, if I have one."[8] An opinion piece by transhumanist Zoltan Istvan even emerged in *The New York Times* in the spring of 2020 during the COVID-19 pandemic, when beaches around the world were closed to surfers, and his opening salvo read as such: "Many surfers like me believe that surfing is more than just a sport; we consider it a way of life. Being in the ocean and riding waves can be ecstatic and spiritual."[9]

For some surfers this love of and engagement with the ocean, with bodily experiencing the sacred center of this performed religious practice, directly motivates them to take part in prosustainable behaviors. As the sociologist Barbara Humberstone found with her investigation of windsurfers, "Social and environmental action, it is argued, emerges from the shared embodied practice; kinetic empathy amongst practitioners of nature-based sports begets action and on occasions this action is in support of social or environmental justice" (2011, 503). This is indeed the case for some surfers, whose kinetic embodied engagement with the ocean can potentially trigger prosustainable behaviors and advocacy-based actions for sustainability. For example, this is seen with Kelly Slater, mentioned above, who started a cradle-to-cradle clothing company that uses hemp, organic cotton, and salvaged fishing long lines to make clothes (OuterKnown[10]) and who has his own signature deck pad (gripping so your back foot does not slide off the surfboard) made from compressed invasive algae that is also biodegradable over time. It is seen with the Chilean big wave surfer and Patagonia brand ambassador Ramon Navarro and his defense of the rivers in Chile threatened by dams. It is seen with environmental educator and social justice advocate Mira Manickam-Shirley, co-founder of Brown Girl Surf, who helps get BIPOC inner-city youth in the San Francisco and Oakland regions into the ocean to wake up their senses and build a connection with the natural world. Their mission statement is a direct example of these leading edges of nature reverencing embodied religions, in this case surfing, responding to climate change and environmental harms: "Brown Girl Surf works to build a more diverse, environmentally reverent, and joyful women's surf culture by increasing access to surfing, cultivating community, amplifying the voices of women of color surfers, and taking care of the earth."[11] It is seen in the thousands of surfers in 2019 who had paddle outs and protested deepwater oil drilling in the pristine Great Australia Bight. Seven-time women's world surfing champion from Australia, Layne Beachley, exclaimed she was involved because "We're basically saying no way Equinor [the Norwegian company behind the drilling campaign]. Australia is not for oil drilling. We need to start looking for more renewable sources for energy."[12]

We see this too with the Irish big wave surfer Fergal Smith, who upon witnessing the Fukushima nuclear disaster in Japan and learning about the impact of CO_2 on the earth's climate and thus on surfing (sea level rise so breaks will disappear), gave up an internationally renowned big wave career, stopped flying, and followed in his parents' footsteps and started his own organic farm in Ireland.[13] In his words, he was "living a dream but not giving anything back. Do I keep traveling and pretend I don't care about the world I live in, or do something about it?" He felt he "needed to go home. I needed to get back to the land and actually do something that is real . . . work that is of real benefit to the community." He now exposes city dwellers and those who have never farmed to the practices and ethos of regenerative farming, largely through a Community Supported Agriculture scheme with his Moy Farm team. Most recently he and his team have taken on reforesting the West Coast of Ireland by planting 12,000 native trees, while still surfing in Ireland when the waves are good. This growth from CO_2 emitting global surfer, to starting an organic farm, to now replanting native trees, speaks to embodied, nature-based practices of sustainable biocultural regeneration, inspired by in this case surfing and a love of nature coupled with a very real concern about climate change. As Smith himself explains in regard to the tree planting, "We want to develop a space to assist people coping with mental health issues and physical challenges, or anyone keen to realign themselves to nature."[14]

And perhaps most famously, at least within surfing subcultures, is the free surfer Dave Rastovich,[15] sponsored by the global surfing powerhouse brand Billabong. Rastovich is such a good surfer that he is paid to just surf, not compete. However, his love for the ocean inspired him in his 20s to become basically a full-time ocean activist. He combines an active meditation practice with being a spokesperson for ocean health, starting the nonprofit advocacy group Surfers for Cetaceans. This group's mission reads: "Surfers for Cetaceans is committed to activating ocean-minded people everywhere to support the conservation and protection of whales, dolphins and marine life. It's through compassion, awareness, education, media and dedicated interventions that we will accomplish this goal. We seek to be a human voice for and defender of cetaceans worldwide."[16] Notice once again the kinship ethics borne of ecological and evolutionary views that for Bron Taylor is a hallmark of both dark green religiosity but also for aquatic nature religions such as surfing.

One of the key actions this group undertook in the past was raising awareness of the pilot whale slaughter held in secret in Taiji, Japan (see also the 2009 documentary *The Cove*). They also undertake active education campaigns with school-aged youth, hold paddle outs at beaches, and are a clearinghouse for educating people about the ecological importance of cetaceans. All of these prosustainable actions are directly motivated by biocentric, posthuman, materialist, religious sensibilities of the bodily kind:

Surfers for Cetaceans acknowledges the many imbalances and issues our oceans face, though focuses on the cetacean issue due to the fact that our global surfing culture has such an intimate human and cetacean relationship. We share the inter species bond of riding waves for the sheer joy of it. The surfers love for dolphins and whales is one born of direct interaction in the ocean waves, and creates a feeling of compassion and kinship with cetaceans.[17]

While this is not the same perspectivist kinship reported by the anthropologist Eduardo Viveiros De Castro (2015) and that forms the basis of many regimes of traditional ecological knowledge, it nonetheless points toward posthumanist assemblages and gives credence to the nomadic posthuman scholar Rosi Braidotti's call for "an ethics of becoming . . . a posthuman ethics for a non-unitary subject [that] proposes an enlarged sense of interconnection between self and others, including the nonhuman or 'earth' others, by removing the obstacle of self-centered individualism" (Braidotti and Veronese 2017, 343–344).

Clearly, the above examples are "ideal type" exemplars, aided by their charisma, fame as paid surfers, class-based post-materialist privilege, and a global surf-mediascape. Yet the development of very clear efforts to make sustainable surf material artifacts, to active campaigning to protect beaches and reefs and waterways, to conscious reduction in flying in order to mitigate climate change, have all gained steam in the last ten to fifteen years. The reasons given by those I investigate above for these include posthuman kinship ethics, new materialist sympoietic becomings with moving liquid energy and ocean creaturely kind, and the desire to move surfing beyond white masculine colonial archetypes. As climate change continues and oceans become more acidic and surf breaks more besieged by changing bio-ecologies of oceanic places, scholars should seek to see if the prosustainable actions shared above, inspired in part by the embodied dramaturgy of surfing-as-religion-stoke, continue to grow and gain more practitioners.

POSTHUMAN EROTIC LOVE FOR
THE NATURAL WORLD

This section builds on the prior, going deeper into what can be called bodily assemblages of the queer kind where bodies enact and perform an ethics of becoming that exists beyond the abstracted, insular individual human of Western thought. This is a becoming-with based on symbiotically and sympoeitically braiding lived, earth-based, nature-spiritual erotic performances with the rest of life, all on a climate-changing home planet. Here erotic is used in the sense of the black feminist theorist Audre Lorde as being responsible to our deepest feelings (2017, 28), especially the affective, embodied feelings of care toward earth and the rest of life (Albrecht 2019). Such care is present in all human animals, regardless of sex or gender (Hultman and Pule 2018).

I approach the content in this section with an understanding that heteronormative binaries, including within religion, shape bodies and thus biocultural modes of bodily being in a way that impacts some bodies violently at the expense of privileging other kinds of bodies. This bodily violence also serves, and at the same time represents, the forces of petroculture and White CisHetero Colonialism that are driving climate changes. Judith Butler helps grasp this bodily violence in her work that astutely explains how bodily performances are reproductive of historical situations (Butler 1988, 521). These situations inscribe power and possibilities (and lack of power and possibilities), and this insight must be extended beyond human animal bodies, too, as part of the posthuman turn, and the needed move beyond petrocultural colonization of all bodies (Gaard 1997). We look, then, for religious bodies in drag, a post-heteronormative materialism where the future of some religious bodies, as explored in this section, will for some be increasingly queered in a move beyond binaries and beyond the bounded ego and cisnormed hetero human body.[18] Here queer is taken to be "both noun and verb, [it is] an ecology that may begin in the experiences and perceptions of non-heterosexual individuals and communities, but is even more importantly one that *calls into question heteronormativity itself* as part of its advocacy around issues of nature and environment" (Mortimer-Sandilands and Erickson 2010, 5; italics added). The verb form of queer is central to the religious-based advocacy around climate change explored below.

These erotic assemblages that include prosustainable action for earth and bio-ecological others in a time of climate change can qualify as quest spiritualities that contain a significant experiential element as articulated earlier by Fuller. Here the quest spirituality is toward experiential, erotic interaction with a sacred more-than-human earth and earth elements and earth processes, moving into posthuman spaces of hoped for bonding, union, and embodied belonging by those performing earth eroticism. Consistent with the dark green humility findings, it is a quest spirituality tethered to a holistic, kinship-based understanding of human bodies as being embedded within evolutionarily evolved sacred ecologies that trigger feelings of awe, humility, respect, joy, and love.[19]

I recognize that longitudinal studies on participants and observers in these emergent post-materialist, posthuman dramaturgical religious performance spaces and assemblages are required to see how self-reported perspectives and behaviors that are a mixed matrix of thought, belief, behavior, action, and performance may change over time in ecologies of place. For example, do emergent posthuman behaviors and perspectives impact both nature and nurture in the meso-level of biocultural sympoietic compost terraworlding? This will require a detailed heterotopology[20] that is beyond the scope of this book.

I also recognize that the two examples in this subsection of performative posthumanism are still within the domain of emerging white, Western, post-materialist embodied and resilient responses to climate changes and earth destruction. As such they may (and most likely do) continue unequal race relations, issues of access, and issues of comfort in environmental circles that reflect different materialist and religious needs and understandings of the natural world (Baugh 2017; Finney 2014) within the context of global warming. This holds true for indigenous peoples, as well, with the journalist Megan Wallace offering a potent and needed critique of the ways ecosex/ecoerotic peoples in white communities need to approach allyship and discussions of nature in their advocacy work (2019).[21] As will be shared, issues of neocolonialism are still present, despite seemingly authentic advocacy in both FuckForForest (FFF) and ecosex groups for total bodily spiritual liberation and flourishing.

I invite readers to explore these above tensions, but recognize here that my choice of these case studies for this subsection is to rather highlight that there are groups of nature-as-sacred practitioners fusing performance, embodiment, and prosustainable behaviors in ways that are far removed from more mainstream "greening of religion" religious groups in post-materialist settings. The former are as worthy of study in academia as are the latter, even if in terms of membership their numbers are drastically lower. As bio-ecologies keep changing, so will religious dramaturgy in queering posthuman ways informed by materialisms, and our scholarship needs to attend to this.

While FFF and EcoSexuals are most likely to be viewed as deviant, especially by those in mainstream and consumer cultures and will most likely suffer continued erasure due to their queering of religion and nature, both, the number of adherents in a religion has no bearing on its "validity," whatever the context of its beliefs and practice. Related is that because these groups operate within capitalist spaces where there is no state support of religious identities and practice, there is a level of salesmanship and self-promotion that is needed in order to share their religious perspectives, their prosustainable views, and their packaged messages. Some may encounter this and view the practitioners as inauthentic, or engaged in "greenwashing." I do not share this analysis based on my research, but I recognize it is one that readers may cultivate based on their own analysis and engagement with FFF and SexEcology. Lastly, the prosustainable behaviors shared below are self-reported. Resource and time constraints make it so I cannot verify them by ethnographic fieldwork of my own, so I invite other scholars to undertake such work.

FuckForForest

When logging into www.fuckforforest.com, the following immediately populates on the screen[22]:

Warning
This site contains sexual and erotic expressions.
The site is created to protect and liberate nature and sexuality.
It contains natural nudity—including graphic sexual and erotic images and sexually explicit language. (. . . hmm?)
It also contains information about how some humans exploits [sic] our planet and suppresses our wild nature.
You should be over 18 years old (some places even 21?!) to enter this site and view the information.
Why??—Because some people in our world still finds [sic] sex offensive and dangerous.
If you are underage or get offended by love, nature or irony, you should exit this site now.

Continuing further means that you understand—and accept responsibility for your own actions.[23]

FFF began in 2004, with two activists, Leona Johansson and Tommy Hol Ellingen. It is now a global movement that solicits and receives pictures, videos, and interviews from members and supporters who are comfortable sharing their nudity, and often, filmed and streamed sexual acts. Claiming to be the "worlds [*sic*] first true ecophilia and eco porn project," they claim that members should not worry as "even if human time on Planet Earth is heading towards its end. FuckforForest will be there with you, entertaining you during these times of horror."[24]

The founders and early members of FFF want to be free—sexually, physically, spiritually, psychologically—from what they perceive to be an earth and body-destroying society. Sporting a mix of tattoos, dreadlocks and dyed hair, piercings, smoking hand rolled cigarettes (possibly joints?), playing music, listening to trance music, wearing used mix-matched clothes, dumpster diving for food, and living in nude-based polyamorous communal spaces, FFF's leaders feel that "because the world is so fucked up," activists are required to free bodies by an ethic of safe, consensual sexual freedom and liberated human nature, both. Mixing Marxist, anti-globalization, post-materialist, and ecocentric epistemologies and discourses, those in FFF engage in "love rituals" to "free body and mind," so that by liberating the values of sexuality there will be no owning of others, and thus no owning of any nonhuman others. In their analysis, such a state of being will lead to a healed human relationship with the planet.[25]

What is key to their form of prosustainable earth-based erotic activism is that "FuckForForest is NOT a commercial erotic website." Rather, according to their website, FFF is "an ecological organization—with a sexy touch. The money you donate to our project—is a donation for us to support nature protection. The humans on the FuckForForest website are not 'actors' paid to do what they do. They are ecological activists and lovers, who really care for nature—not ashamed of showing YOU how nature created them. So FuckForForest is—ecology and sexual/body liberation. ALL IN ONE!"[26]

According to their website and the below referenced documentary on them, FFF is a nonprofit registered in Norway. In legal theory this means all their expenses have to be accounted for and must be transparent. FFF claims to basically be a clearinghouse of funds: they market themselves as accepting donations by their website through the practice of sexual liberation via self-produced and solicited photos and videos, the content of which is used to attract viewers and for-pay followers. Viewers and followers are solicited for donations—the pro-environmental messaging of FFF's website and the natural settings of the pictures and videos are catered toward those concerned with environmental issues. When enough money is in hand, FFF then solicits requests for donations from earth activists around the world and the FFF leadership collective decides who receives funds. According to their website, they target smaller groups who need financial support with the focus of the money raised supporting the protection of old-growth rainforests, and replanting of logged areas, in Central and South America. Projects funded from 2006 through 2017 include reforestation and forest protection in Costa Rica, Ecuador, Peru, Brazil, Mexico, and Slovakia.[27]

The cover for Michal Marczak's documentary on FFF, which as of summer 2020 was also the picture banner on FFF's Facebook page (see Figures 7.2 and 7.3), shows scantily clothed and topless indigenous peoples surrounded by forest clear cuts, while FFF members, who are also nude, engage in erotic coupling in a few healthy trees and dance around a campfire, observed by the indigenous peoples. Tensions between FFF members and their exotic gaze, and indigenous mores around nudity and sex (Prager 2017), ironically come to a collision in the documentary. In one of the film's closing scenes the founders of FFF journey by boat seven days deep into

Figure 7.2   The Fundraising Image Used as the Banner for FFF's Facebook Page as of June, 2020. *Source*: Used with permission by FFF

Figure 7.3 Branding/Marketing Poster Created by FFF. Note the posthuman, esoteric, nature reverencing depictions of bodies and bodies performing what FFF claims are sacred sex acts. *Source*: Used with permission by FFF

the Amazon and attempt to help indigenous peoples there buy land to help protect it. At a meeting organized by the indigenous leader who invited FFF (he journeyed to Quito where he sent out a call for help via internet, and this call eventually found FFF), many in the audience claimed that FFF members were full of European lies and that they prostituted children. Claiming via a translator that "we know how to care for our trees, as we have been doing it for 500 years," those present said their biggest need is jobs to get out of poverty, and FFF is unable to help with that. After being invited to leave, the camera pans to a chainsaw salesman who is making a pitch to the same assemblage about the benefits and comfort of the newest model chainsaw on the market (Marczak 2012). The layers of culture clash, different views of nudity and bodies, different understanding of traditional ecological knowledge and market forces, and the crushed idealism of the FFF members are on stunning display in this sequence. Despite this failure, the group has nonetheless been able to organize and support via donations the protection of multiple acres and regions of rainforest, helping protect these ecologies of place for their bio-ecological functions.

FFF's approach to a queering past the human is experiential and tactical. On the experiential side, they advocate for the liberation of our bodies—we are "organic," and thus our sexual emissions in all their forms are natural. As shared in the documentary and on their website, for them sexuality is the life-creating force of the universe and this sexual life force used to be celebrated in pagan cultures. For them, it is time to reclaim this worship of the sacred erotic from a modern capitalistic culture that has made such erotic actions deviant, taboo, and prohibited, often by fines and penal repercussions. To do this members often engage in same-sex, pansexual, and group-sex performances, often at house parties where the audience are asked to donate monies to their nonprofit. This relates to the tactics for FFF of earth care: as they say in an interview, "It would be stupid to just affect the people who already get it. It was also about the ecological part of it: many people who go to ecological websites are already interested in ecology. So how do we create an alternative that takes the people in who need information about an alternative way of thinking to maybe change their values?"[28]

Inspired by participating in the international "SlutWalk," which raises awareness about how women who dress "provocatively" often get raped and are shamed, while the violent perpetrators often do not go to jail because the victim was dressed "sluttily," FFF created the slogan "Nature is a Slut." For them this is a phrase of liberation, of challenging the commodification of sex and nature both, and of celebrating the fecundity of nature's creations.[29] For FFF, the absurdity of a fossil fuel-based consumer culture that represses genuine bodily relations, and that is thus killing the earth, is the true crime. As they explain, "Because we actually think a car is more dangerous than a naked person or a couple making love, we have no problem going out in the streets to challenge that car—and its moral values in society."[30] This challenge is performative, embodied, lived, expressed, and based on very clear religio-spiritual connections to the more-than-human world, connections they call "surreal, recycled, shamanistic, chaotic" (quoted in Măntescu 2016, 15) and that dramaturgically get enacted at sex performances and shows. For the FFF collective, this connection to and love of nature, as well as their connection to and love of bodies, is their inspiration to engage in embodied, prosustainable behaviors: "Without human consciousness on the planet you have no rights so the values and morals of our society have to change. So making the world more connected to nature for example is the most important part of FFF in the longer term."[31] For this nature spirituality group operating in posthuman materialist spaces, where making love in the natural world leads to a bonding with sexual partners and the natural world, both, protecting old-growth rainforests is their pathway of bringing their post-materialist resilient religious ideals in an era of climate changes into practice.

SexEcology

Annie Sprinkle and Elizabeth Stephens are the married couple behind (see Figure 7.4) the SexEcology movement. Sprinkle holds a PhD and is an artist, while Stephens is also an artist and teaches college classes. That these two have founded the EcoSex movement is not by happenstance. Sprinkle herself[32] has a history fusing performance art, sex work, and burlesque performance (Williams 1993; Montano et al. 1989); while Stephens grew up closeted in Appalachia and after coming out in urban-punk gay enclaves and seeing the damaging stereotypes educated liberal gays in these areas had of Appalachian communities, now defines herself as "a proud tree-hugging, herb loving, animal cuddling ecosexual and I will never return to any closet again."[33] Both merge their feminist backgrounds with concern about the destruction of nature so that EcoSex is in part an ecofeminist movement working on dismantling the interlocking oppressions of female/nonmale bodies and agencies and the natural world (Plumwood 1993).

However, in Sprinkle and Stephens' alchemy, EcoSex is an effort "trying to make the environmental movement more sexy, fun, and diverse, and tap into unexplored audiences."[34] These are similar goals as FFF, and like FFF, the post-materialist entryway into more fun for EcoSex is through erotic embodiment based on love for the natural world, and thus proactive efforts to protect nature's diversity. To this end they both self-label as EcoSexuals and have scribed a six-part EcoSex Manifesto:

1) We are the ecosexuals. . . . In order to create a more mutual and sustainable relationships [*sic*] with the Earth, we collaborate with nature. We treat the Earth with kindness, respect and affection.
2) We make love with the earth. We are aquaphiles, teraphiles, pyrophiles and aerophiles. We shamelessly hug trees, massage the earth with our feet, and talk erotically to plants. . . . We make love with the Earth through our senses. We celebrate our E-spots. We are very dirty.
3) We are a rapidly growing global community of ecosexuals. This community includes artists, academics, sex workers, sexologists, healers, environmental activists, nature fetishists, gardeners, business people, therapists, lawyers, peace activists, eco-feminists, scientist, educators, (r) evolutionaries, critters and other entities from diverse walks of life. . . . As consumers we aim to buy less. When we must, we buy green, organic, and local . . . we connect and empathize with nature.

Figure 7.4   Sprinkle, Left, and Stephens, Right, Taken from Their Press Kit at https://th eecosexuals.ucsc.edu/press-photos/. *Source*: Photo credit to Julian Cash; image credit to Beth Stephens and Annie Sprinkle.

4) We are ecosexual activists. We will save the mountains, waters and skies by any means necessary, especially through love, joy and our powers of seduction. We will stop the rape, abuse and the poisoning of the Earth. We do not condone the use of violence, although we recognize that some ecosexuals may choose to fight those most guilty for destroying the Earth with public disobedience, anarchist and radical environmental activist strategies.

5) Ecosexual is an identity. . . . Ecosexuals can be GLBTQI, heterosexual, asexual, and/or Other. . . . We are polymorphous and pollen-amorous. We educate people about ecosex culture, community and practices. We hold these truths to be self-evident; that we are all part of, not separate from, nature. Thus all sex is ecosex.

6) The ecosex pledge. I promise to love, honor and cherish you Earth, until death brings us closer together forever.[35]

This pledge covers many of the themes and arguments of this chapter. Stephens and Sprinkle in this manifesto move human identity into an embodied posthuman materialist space; and they claim that their love of earth should inspire prosustainable, resilient behaviors related to dietary choice and habits of consumption by EcoSexuals. They are also clear that education and activism are central to an EcoSex identity, where for them a key part of this identity is in service to protecting the natural world. And key to an EcoSex identity is an erotic, embodied, love-making with the earth and its elements. Their merging of "ecological, cyborg and sex-positive feminism [with] activism addressing sex work, queer sexuality/theory and [imbuing] it all with a post-modern sense of play, paradox, and irony [creates a] radical re-articulation of desire's directionality [and] opens up a breadth of future possibilities" (Weadick 2013, 90). And what is one of the more key pathways of and to human embodiment than our physical, and especially sexual, desires? And what can be more desirable for some, seeing planetary metrics, than creating a more sustainable future that adapts to climate change and protects the flora and fauna of the earth? And doing so out of religious-based love for an earth perceived as sacred beloved?

Sprinkle and Stephens share on their website 25 ways ecosexuals can make love to nature. This list is divided into four element-based ecosexual pathways where for intimate ecosexual actions one can practice with their bodies teraphilia (e.g., vegetable dildos, rock fetish, making love on grass or moss, mud wrestling); for air, one can practice aerophilia (e.g., cloud sex, erotic asphyxiation, ecstatic breathing); for fire, one can practice pyrophilia (e.g., nude sun bathing, fire walking, volcano fetish); and for water, one can practice aquaphilia (e.g., surfing, waterfall play, rainplay, skinny dipping).[36] These identities and practices are fluid, so as one transitions from one elemental practice to another the bodily practices reflect this—for example, the practice between aquaphilia and pyrophilia is a steam bath.

Their efforts also include hosting EcoSex weddings; offering EcoSex walks so participants can better know and engage—erotically—with their local habitats; and using film as a way to educate viewers about pressing environmental issues. Stephens and Sprinkle have over a series of years been married to the Moon, to the Appalachian Mountains, and other natural features by ritualized dramaturgical wedding performances where vows are exchanged with these natural elements. As they write,

> With these weddings we wanted to raise the following questions. Can the Earth, with its Mountains and orbiting Moon feel our love? How can humans join together with nonhumans to create mutually sustainable relationships and communities that flourish in the face of extinction? Can ecosexual unions be consensual? Is anthropomorphing nonhuman entities really helping, or is it another form of violence or both? We have no conclusive answers to these questions. What we hope is that through the embodied offerings of performance, poetry, music and art we are creating rituals that open human hearts to all life forms and perhaps this love helps the Earth continue to flourish. Many of our collaborators and wedding guests told us that they have taken their vows seriously and feel

more connected with the Moon, Mountains and Earth since the wedding. Some have said that they have changed their habits to more fully embrace sustainability. We certainly have changed our daily habits, being more aware of water usage, gas, recycling and other sustainability practices. . . . Although we believe that we are not separate from the Earth, or nature, as some religious institutions and social constructs claim, our weddings provide a dynamic form to unite us all. We believe that the act of making vows and sharing the benefits of love, caring and responsibility that the institution of marriage claims to represent and guarantee must be extended beyond humans. (2012, 64–65)

I quote Sprinkle and Stephens at length as this passage contains a mirror of one of the key research questions of my book and the focus of this chapter. It also provides evidence of a posthuman material embeddedness and embodiment with nonhuman earth, and one that is strategically ritualized, where those present in the ritual report an increased likelihood of engaging in prosustainable behaviors. Yet such queered earth/sex embodiment and ritualization is found mainly in post-materialist spaces, buttressed by a critique of capitalism and patriarchy informed by Western discourses and access to basic material needs being met.

In regard to education, Stephens and Sprinkle released a 2017 documentary turned into an 2019 Amazon.com road-trip based documentary series, *Water Makes us Wet—An Ecosexual Adventure*, where they travel the United States investigating pressing issues related to water such as water pollution and looming water shortages caused by poor water management and climate change. They also released a 2013 documentary, *Goodbye Gauley Mountain: An Ecosexual Love Story* about mountain top removal for coal in West Virginia, the ecology of place where Stephens grew up. For example, in episode one of *Water Makes us Wet* the narrator Sandy Stone as voice of the earth explains how Stephens and Sprinkle see her (i.e., the earth) as their lover, where they are happy in earth's waters, some of which they have even ritually married. The episode defines ecosexuality as "an expanded form of sexuality that imagines sex as an ecology that extends beyond the physical body" (2019), so that to be an ecosexual means someone is "a person who imagines the Earth as their lover, and finds nature (human and non-human) sensual and erotic' (2019).[37]

This narrative script is an anthropomorphism of earth, presented as witness to Sprinkle and Stephens' road trip. Stephens and Sprinkle weave in a bioregional sensibility, where earth exclaims, "While humans created maps to conquer my surface, I've always divided my waters by watershed boundaries. My wet spots were the original mapping technique" (2019). Throughout the episode water is utilized as metaphor, but presented as the foundation of life that connects all and creates all, that is soothing but also dangerous and violent, as nature can "be a cruel lover with no safe word" (2019). They also undergo a divine feminine card reading in the shape of a *mandala*; interview a variety of scientists and activists and artists; and engage in a series of educational, humorous, and poignant sensory bodily based communications with other humans, seals, streams, rivers, fish, and the ocean.[38] Throughout viewers are educated about the impacts of climate change on ocean and water health; the impacts and causes of water pollution; and the deleterious impact of the private water bottling industry on wildlife and groundwater. There is even a scene where they wash away the "ecosins" of a friend in a swimming pool!

The above embodied earth-based religio-spiritual ritual bonding with the elements championed by EcoSexuals, and as performed by Stephens and Sprinkle, have an uncanny bond with absurd theater. As the professor of Drama Ralph Yarrow explains, "The 'absurd' is a *declaration* of crisis. . . .The absurd in theatre is a machine to produce the explosion of all known categories, systems, limits, modes, forms and frames, and leave you in the resultant void/vortex/vacuum. So the absurd is also a kind of sacred wound, a tearing apart or splitting open, in and from which the freedom to make anew is engendered" (2007, 60–61, italics in original). In many ways, both FFF and EcoSexuals undertake precisely this movement, both internally and also in terms of using their bodies-in-public-erotic-relations-with-earth-as-activism. The crisis being declared is multifold: a crisis of economics, of politics, of flawed conceptions of our bodies and what sex and the erotic is, of climate change, and of religion. Ritualized, liberatory sexual engagement with the "other"—the natural world—is an absurd pathway to explode the perceived false categories of human, other, nature, and self. Rather, embodied eroticism with a sacred nature is undertaken precisely by those in these movements to heal the wounds that have caused, and that are manifest in us, by the interconnected crises pointed to by FFF and EcoSexuals in their post-material analysis of mainstream culture.

By getting back to our animal bodies, and proactively eroticizing them and the other bodies in our midst, these movements create a posthuman materialism that invites fellow adherents and participants to make anew their relations with sacred nature, in part based on the concept of fun. Part of this erotic making anew includes stated commitments for engaging in prosustainable behaviors. That these particular post-materialist moves may engage in cultural appropriation and may not be of interest to certain communities, especially indigenous or of color, or to those more comfortable with mainstream religions, must be factored into how they operate as well.

Despite the concluding sentence of the above paragraph, seeing FFF and EcoSex as forms of nature-based religions helps add a needed level of analysis in understanding both. It is clear that the materialist, embodied, posthuman turn this book is advocating for as its analytical lens also helps to understand the nascent growth of erotic ecosexualities. This foundation of sensual, embodied engagement with nature, where one's identity may even extend beyond the physical body, is evidence of potential new ways of being human, at least in post-materialist Global North contexts and ecologies of place, where erotic interaction with and loving care for the earth motivates religious practitioners to grapple with creating prosustainable and post-petroculture lifeways.[39]

From a religious studies perspective it is clear that those in surfing and EcoSex/ecoerotic subcultures fit within the dark green religion schema articulated by religion and nature scholar Bron Taylor. This form of religion sees nature as sacred, and Taylor's definition of four interlocking types of dark green religion—spiritual animism, Gaian spirituality, naturalistic animism, Gaian naturalism (2010)—helped create the survey matrix he, Jen Wright, and I distributed. The results of that survey and the case studies presented in this chapter suggest that these and other religions-at-the-margins are forms of "emergent Earth-centered religious production on a planet undergoing a variety of anthropogenic environmental tipping points" where such production "in regards to sustainability is occurring and will continue to occur as environmental degradation continues in the years and decades to come" (LeVasseur 2020, 88). The challenge on the academy, and especially religious studies, is to see how studies such as these can possibly, or will actually contribute to resilient transformations, if at all. Or, if not to that, then to a compassionate movement into likely petroculture collapse and radical social, economic, and environmental impoverishment; and even to potential species extinction if predicted worst case (statistically possible but highly and extremely unlikely) climate scenarios ever come to pass.

How successful these religions-at-the-margins actually are in creating adaptive and resilient material biocultures requires further studies by those in academia working on these issues. However, based on self-reported behavior change, the evidence does suggest initial and reasoned efforts of such prosustainable behaviors in post-materialist, petroculture dwelling spaces. If able to be sustained over time, or more, inspire such behavior in new adherents and dramaturgical practitioners, then they do present hints at measurable adaptive bioculture materialities in post-material spaces that are inspired by posthuman identities. This raises important questions about the possibility of solutions to climate triage emerging from late capitalist petrocultures. Or if these behavior changes are so miniscule compared to what is required to avert full catastrophe (which may likely be too late at this point), or at least redesign resilient adaptivity, then are they just the privileged whimsy of those wanting to liberate their bodies via the vehicle of a professed earth care as the world burns? As shared earlier, FFF hints at this possibility in their most recent iteration of their homepage. And if by some luck of human effort and redesign that true adaptivity is to be achieved, would it not make more sense to allow ecosystems people their autonomy by helping protect their bio-ecologies from violent resource extraction and dispossession by omnivores, and to then craft policies and economic patterns based on the knowledge base of ecosystem peoples attuned to bio-ecological places? Yet is this possible without some form of appropriation, neocolonialism, romanticizing, and power-over dynamics, given the last 500 years of globalization and Settler Colonialism? And is it feasible to expect youth in ecosystem cultures to want to remain in them, when other opportunities beckon, as I witnessed in Ladakh? I do not have answers to these pertinent questions but raise them nonetheless as these are the types of questions our field needs to be addressing to remain relevant in a time of rapid climate changes.

NOTES

1. Post-materialist here means post-industrial; or living in an economic regime where there are enough jobs and resources that basic, material, quotidian needs of food, clothing, and shelter are easily met and therefore other needs are able to be pursued, like shopping, going on vacation, joining a gym, undertaking advanced education, and others. That is, this is not to be confused with moving beyond an agental, materialist understanding of the observable earth and universe that is articulated in this book.

2. "When we used Pearson's correlation to find correlations among and make inferences about variables and perceptions surveyed, our hypotheses were found valid at levels of statistical significance: that dark green religion and environmental humility positively correlate; that dark green religion and environmental humility positively correlate with prosustainable behaviors; that respondents who already have interest in environmental issues had higher levels of dark green religion and environmental humility and a stronger relationship between these and prosustainable behaviors; and that anthropocentric and monotheistic perceptions and correlated views of humility do not foster, or do not foster nearly as much, prosustainable behaviors as do people reporting Gaian/pantheistic/animist/Organicist (subsets of dark green religion) religious perceptions" (Taylor et al. 2020). As we point out in that article, the question of which comes first: religious identity and then prosustainable behaviors, or vice versa, is an open one. The relations between these dynamics are rich and deserve more study across multiple fields, especially longitudinally, although insights from adaptive management and traditional ecological knowledge suggest that what fosters prosustainable behaviors is that a resource crisis is often experienced that is then followed by a change in cultural values and practices that better protect and steward that resource (Berkes 1999). The reader should keep this in mind later in this chapter when I briefly explore post-materialist religious bodily dramaturgy and prosustainable behaviors—if Berkes' insights hold, these behaviors and religious identities are

largely a response to perceived and felt bio-ecosystem decline. However from an emic perspective these behaviors are valid, affective, posthuman, queering-of-bodily identities and performances via emerging religious embodiment and practices that are centered around nature-as-sacred perceptions and value systems.

3. Here sustainable is meant in the sense of being able to be sustained longer than dominant fossil fuel-based choices related to energy, transport, and food, but is not to be mistaken for regenerative. This is a different question to ask, and harder to measure, as most of us are not conversant with what regenerative systems of agriculture, transit, and energy would resemble, let alone how these would be built and adopted. As Wes Jackson explains, these are very real challenges we need to figure out. To be regenerative means we will have to be the first species to say "no" to the 3.45 billion-year imperative, while creating just, equitable systems of thriving built upon regenerative agriculture, energy, and transit. This implies governance and policies in support of such regeneration, and citizenry cognizant of what regenerative choices in these domains could be, and why we need them. We are currently nowhere near to having such a politics, in no small part because we do not have such an educational system that teaches and embodies regenerative solutions to adaptive, sustainable living in ecologies of place.

4. I thank Louise Doire in The College of Charleston's Religious Studies Department for this framing who, upon hearing me describe this book project, said "You're using climate change to do hermeneutics."

5. Which is not to say that such forms of privilege should not be continuously lessened by those benefiting from them.

6. I thank Steve Pezman, the editor of the fantastic *The Surfer's Journal,* for this insight on the difference between wave riding, and surfing. This signals to a transhuman future for wave riding.

7. To be clear, many surfers would not self-classify as being part of a quest spirituality. This may especially hold true for evangelical Christian surfers who would most likely see this as a type of idolatry.

8. https://www huffpost.com/entry/religion-of-surfing_n_5617472 Accessed May 11, 2020.

9. https://www.nytimes.com/2020/05/01/opinion/surfing-coronavirus-quarantine .html Accessed May 12, 2020.

10. https://www.outerknown.com/ Accessed May 11, 2020. OuterKnown's mission statement is to "transform the way we outfit the world by inspiring an industry to reimagine design and embrace circularity," while the company's vision is, "A future where clothing contributes to a world where people and planet prosper." To learn more about the sustainability operations and innovations championed by Slater and his company, see https://www.outerknown.com/pages/sustainability.

11. https://www.browngirlsurf.com/our-mission Accessed May 11, 2020.

12. https://www.cbsnews.com/news/surfers-protest-oil-drilling-in-the-great-aust ralian-bight/ Accessed May 11, 2020. This can only be a quick "surface" level summary—for example, it is assumed many of the surfers and water protectors who paddled out did so because of more selfish reasons: they did not want pollution in their recreation areas. Yet, this assumes such recreation is not tethered to spiritual

experiences—it is as possible that many protested precisely because they experience the ocean as a sacred place and do not want it, and the lifeforms therein, to be damaged and polluted.

13. For a TEDx talk by Fergal Smith, from where these quotes originate, see https://www,.youtube.com/watch?v=kmyuF70KRG8 Accessed May 11, 2020.

14. https://www.redbull.com/us-en/fergus-smith-surfing-eco-farming-moy-hill Accessed May 11, 2020.

15. Rastovich and his partner purchased an overgrazed farm outside Byron Bay, Australia and like Fergal Smith have planted over 7,000 native trees and practice permaculture on the land, nurturing it back to health. For more on this aspect if his ecological civics, see https://www.surfer.com/features/in-byron-bay-lauren-hill-and -dave-rastovich-reimagine-abundance/ Accessed May 18,2020.

16. https://www.s4cglobal.org/ Accessed May 18, 2020.

17. https://www.s4cglobal.org/about.php Accesses May 18, 2020.

18. It is important to point out that the analysis implied by such looking for applies to those of us in dominant petrocultures and cultural manifestations of, including the academy. It is a form of neocolonialism to universalize and project prescriptions for Global North sexual politics onto indigenous, empty belly, ecological refugee, and peasant/agrarian communities. This does not mean that very real suffering and violence does not occur in these other communities around sexual and gender identities and performativities. Solidarity to liberate from and actively dismantle queerphobia wherever it occurs is needed; offering a prescriptive pathway to such solidarity, let alone on a global level, is not the point or goal of this chapter.

19. It is very much the case that surfing and other outdoor activities mentioned in the prior section can easily be analyzed by these same epistemic lenses. For heuristic reasons I kept this section to FFF and SexEcology, to highlight that their explicit way of queering bodies is through sex practices and gender identities. This specific pathway to the erotic is not as present in surfing and other outdoor nature-as-sacred identities and practices.

20. Heterotopias is a concept developed by Michel Foucault and are "real places—places that do exist and that are formed in the very founding of society—which are something like counter-sites . . . in which the real sites, all the other real sites that can be found within the culture, are simultaneously represented, contested and inverted" (quoted in Rebentisch 2013, 142). FFF and Ecosex are found within White CisHetero Settler Petroculture Colonialism, and seek to contest and invert such culture. Members do so via nature-as-sacred religiosities that inform bodily dramaturgies that include stated efforts of prosustainable behaviors. However, the aesthetics philosopher Julian Rebentisch points out that political charisma in an age of mass media has to prove itself against other charismatics in order to communicatively interact with a public audience (ibid, 158). Similarly, then, for a religious charismatic to emerge in an age of mass media will require dramaturgy that appeals to the masses. For this reason, FFF and Ecosex will most likely remain on the margins, despite any inherent charisma of their founders—most people will probably not view sex, love, the erotic, and the body in ways that those in FFF and Ecosex do, even if they do hold nature-as-sacred worldviews. This is why there will most likely continue to be more surfers

than ecosexuals. And why the phenomena outlined in the case study of chapter 6, of Buddhist monks and ice stupas in Ladakh, will have the biggest potential prosustainable impacts: the religious and thus cultural charisma of the monks are still held in high esteem, even in an age of mass media.

21. The article contains important insights from an interview with Native American decolonial sexualities scholar Kim TallBear. Williams writes, "As you can imagine, ecoerotics resonate with white, dreadlocked, neo-hippies looking to reconnect with nature, making the appropriation of indigenous practices a risk." After which she shares that, "TallBear has previously warned against the 'appropriation of Native American knowledges and motifs to the ecosexual ceremonial and artistic repertoire.' She also, in conversation with *Teen Vogue*, pointed out that the practices and perspectives of some ecophiles are the epitome of white nonsense: 'Ecosexuality is not going to appeal to most indigenous people . . . I teach it in my classes and my students are viscerally like, 'This is weird, self-indulgent white people' . . . When people talk about the Anthropocene they typically say, 'We as a species are now coming to realise that we have to stop putting humans at the top of the hierarchy. Other beings have agency,' and I m like, 'No, it's not we who are just now having this revelation; it's a bunch of white guys.'"

22. It should be noted that FFF is now behind a paywall, but according to its founders and webhosts not for greenwashing and self-aggrandizement/enrichment purposes. Rather, when FFF began their web-based activism there was minimal porn content on the internet—what was there was largely for pay. Thus when they started, they asked for donations through their free site of mostly self-produced at minimal cost porn and pictures of nude bodies. Given the rise of Pornhub and other free porn sites, FFF now has a paywall as a way to guarantee the raising of money. That is, their fundraising strategy is that those committed to their cause will pay because of the cause, not just to view porn, given that porn is now everywhere on the internet versus when FFF began. For FFF the paywall is a way to weed out porn voyeurs and to actually strengthen their community of participants and supporters by show of pledged monetary commitment.

23. www.fuckforforest.com Accessed May 23, 2020

24. www.fuckforforest.com/page/about/ Accessed May 23, 2020

25. Quotes from and analysis based on the 2012 documentary *Fuck For Forest*, directed by Michal Marczak.

26. https://www.fuckforforest.com/page/about/ Accessed May 23, 2020.

27. https://www.fuckforforest.com/page/projects/ Accessed May 23, 2020.

28. https://mastazine.net/?q=handbook/interviews/fff-still-fucking-forest Accessed May 23, 2020.

29. For an analysis of the reclaiming of the term "slut," see: https://www.newstatesman.com/politics/feminism/2017/09/amber-roses-slutwalk-controversial-feminist-movement-still-relevant. Accessed May 23, 2020.

30. https://vamosfalardesexualidadeblog.wordpress.com/2020/01/22/interviewing-fuck-for-the-forest/ Accessed May 23, 2020.

31. ps://mastazine.net/?q=handbook/interviews/fff-still-fucking-forest Accessed May 23, 2020.

32. Sprinkle self-identifies on her website anniesprinkle.org(asm) as "a feminist porn activist, ecosexual, and radical sex educator." http://anniesprinkle.org/projects/current-projects/my-life-and-work/ Accessed May 23, 2020.

33. http://elizabethstephens.org/eco-sexual-history/ Accessed May 23, 2020.

34. https://www.dazeddigital.com/life-culture/article/44391/1/ecosexuals-eco-porn-sex-environmentalism-elizabeth-stephens-annie-sprinkle Accessed May 23, 2020.

35. http://sexecology.org/research-writing/ecosex-manifesto/ Accessed May 23, 2020.

36. http://sexecology.org/research-writing/charts-and-graphs/ Accessed May 23, 2020.

37. This scheme allows surfing to be an ecosexual practice if a surfer identifies as such. I am personally curious if members of the FFF collective and Stephens and Sprinkle have interacted in any way, given the strong overlap in their erotic practices, posthuman views, and desire to protect the earth.

38. For an analysis of the documentary and its role in queering ecological imaginations, but that points toward alterity rather than the reciprocity of Stephens and Sprinkle, see Nuruan Tola (2019).

39. To be clear, some of these developments and engagements are found/are emerging in religious dramaturgical systems and biocultural places whose sources are not white EuroAmerican. These prosustainable developments—for example, hybrid and syncretic bodily practices as response to settler petroculture in Santo Daime, or nonwhite pagan spiritualities—belong within the broad religious biocultural landscape I am discussing here. However I lack the familiarity with them to discuss them in this book.

Conclusion

I conclude by returning to a key underlying concern that has motivated this book, with this concern articulated in the words of the political scientist Jedediah Purdy, who writes "[C]limate change, mass extinction, ocean acidification, soil exhaustion—the world that may be coming to destroy us is also the world we have made. And of course it isn't simply 'we'—it's the effects some of us are having on the planet, unequally visited on others, through the medium of the world itself, its floods and droughts and killing heat" (2019, 21). And then here, his "money quote": "The natural world, the land, is the thing you can always tell lies about, because it doesn't answer—until the time you can't lie about it anymore, because it's too late" (2019, 21). Here, then, is my concern, as I have written these pages and seen more climate data points emerge over the course of crafting them, and more lack of action at the scales needed to adapt to a warmer future. My concern is that our lies, including those woven into and perpetuated by the academy, are coming home to roost. My concern is that the agental world, the land, the material places within which we dwell, the material medium of the world itself, is in its way saying, "No more." No more CO2. No more methane. No more deforestation. No more extinctions. No more rampant consumption and myths of perpetual economic growth. No more exponential human population growth. It is over, from the unequal destruction beginning with the Agricultural Revolution through the soon-to-end flourishing of petroculture: no more. The Swedish teen climate activist, Greta Thunberg, summed this up eloquently in 2019 when she said to a gathered audience of climate leaders: "Entire ecosystems are collapsing. We are in the beginning of a mass extinction, and all you can talk about is money, and fairy tales of eternal economic growth. How dare you!"[1] No more, indeed.

This concern about how we continue to lie to ourselves about our separateness from and superiority to the natural world within which we always reside informed the two interconnected research questions of this book. The questions I have asked in this book are a response to such lies, to a short sighted and willful silencing of the rest of life, found deep even within the halls of the academy. Trained into such silencing my whole life through my educational journey, in this book I have attempted to find my voice and to create a platform for the voices of agental-others on our shared planet to speak. In the process I have asked: as bio-material beings, how will our biocultures shift as the ecologies of place within which we dwell are transformed by climate change?; and given these shifts and transformations, then how will the bodily dramaturgical practice of religion change as ecologies of place change?

Implied in both questions is that the material biogeochemical and bio-ecological preconditions upon which religions at the meso-level have bioculturally evolved no longer exist. The Holocene epoch of our shared planetary history (including religious history) is done. Geologically speaking, the last time there was this much CO2 in our atmospheric commons there were no ice caps and palm trees were found at the poles. That is the planet religion will be embodied on in the centuries to come. This future will be upon us sooner than we think, especially in the context of geological time, the lies we still tell ourselves be damned.

Also implied in my guiding research questions is a new (and for some biocultures, ongoing and/or old) way of materially understanding the human animal on our evolving planet that we co-create and which co-creates us. Toward the beginning of the book I shared an insight from Krizia Nardini and wish to do so here again at the end. When thinking about agents, agency, bodily affects and planetary effects, we need to understand that "Ontology is conceptualized as a rhizomatic network of becoming(s) or material-discursive intra-actions. What we are witnessing is a reality with no copies and no originals, in which everything becomes intransitively, at various speeds and intensities, interconnected with other processes of transformation(s). Therefore, within this radical immanent ontology of becoming, there is no room for structural dualisms or representationalist instances" (2014, 21). The implications of this are that we are radically contingent beings dependent upon the effects and agency of literally the rest of material life—both its abiotic (effects, but also potential for agency) and biotic (agental) factors. This stretches as far out as the imploding hydrogen of our sun that powers the aquatic and terrestrial trophic food pyramids upon which we depend for our caloric sustenance; to the bacteria that line our intestinal tracts. And this co-becoming is always the material reality of communicative dwelling within which we reside.

There is no copy of this reality, no distancing from it, no silencing it (despite efforts)—it is as Nardini says a radical and immanent ontology. A literal ground of being, upon which we always walk and that nurtures us and shapes us by rhizomic networks of material-discursive intra-actions. It is upon this same ground—the bio-ecological places this book has explored--that our bodies enact and perform religion.

To answer the above two questions that motivated this book, informed as they are by the rapid climate warming of bio-ecological material contexts, I utilized a variety of thoughtways. New materialist theory allowed for an understanding of other nonhuman bodies, bio-ecological places, and agental biogeochemical flows as having performative impacts upon our biocultural bodies. Material feminism added to this insights related to intersubjective performativity and communication between bodies, and a need to recognize the gendered elements of climate change. An underlying current of ecowomanism highlighted how logics of domination are always at play, especially when conceiving of and regulating (violently) certain human bodies, and the nonhuman bodies and bio-ecological places within which we reside. This is especially true for the violent machinations of petroculture and workings of carbon democracy.

Meanwhile posthumanism challenged Cartesian dualisms, assumptions of human superiority, and the sense of radical separateness from the places within which we always dwell and that we always shape and allowed us to rethink our own animal bodies in ways beyond such fallacious binaries. Ecological animism unveiled the world, and the agental, material others with whom we share it, as alive, populated by other types and bodies of agency and relational becoming. In such animist spaces a performance studies lens led to a dramaturgical lens to investigate religious bodily performances in ecologies of place. Queer theory then facilitated a move beyond dangerous and limiting binaries to instead grasp entanglement and interconnection. Such interconnection is even eroticized, in understanding "that on this planet there is a foundational erotic attraction between all bodies" based on material "phenomena of exchange, of relationship, and of contact" (Weber 2017, 17, 18).

And lastly Indigenous Knowledge provided evidence that there are indeed other ontologies and epistemologies of place and of the agental beings in those places; that we can move beyond petroculture; and that the academy is rife with still ongoing colonial power dynamics of silencing that extend into how we conceive of human bodies (as isolated) in bio-ecological places (as inert and as background). Rather, for many Indigenous biocultures there is still ongoing engagement in "an interspecies community or networked set of social-biological relations [with] living beings that are both material and *immaterial*" (TallBear 2011, italics in original). Here it is important to recognize, as does this book, that "signs are not exclusively human affairs. All living beings sign. We humans are therefore at home with the multitude of semiotic life. Our exceptional status is not the walled compound we thought we once inhabited" (Kohn 2013, 42).

I utilized an evolutionary epistemology, as well, to think about biocultures at the meso-level, where such biocultures unfold within the larger evolutionary scaffolding of micro- and macro-level sympoietic terraworlding. From the perspective of object-oriented ontology, "There is no exit from this situation. Thus the time of hyperobjects [of which global warming is one such] is a time of sincerity: a time in which it is impossible to achieve a final distance toward the world. But for this very reason, it is also a time of irony. We realize that nonhuman entities exist that are incomparably more vast and powerful than we are, and that our reality is caught in them" (Morton 2013, 130). And consistent with an evolutionary epistemology, to answer the questions of this book we have journeyed to a multitude of material bio-ecological places. In these ecologies of place humans at this meso-level dramaturgically perform religion with their bodies as part of adaptive or maladaptive (especially in petrocultures) biocultural dwelling, where such bodily religious performance does not distance itself from the world, but acts instead in recognition that we are caught, precariously, in the world. A world that is alive, and that acts and communicates.

At Standing Rock we encountered indigenous youth and indigenous elders inspired to protect this living world by performing ceremonial-based resistance to petroculture, where such resistance was tethered to and informed by indigenous ontologies and conceptions of sacred places. These responses are part of even larger and centuries long movements of resistance and transformation to violent dispossessions enacted by the (il)logics of White CisHetero Colonialism.

We journeyed to the Himalayas and Gangotri, source of two of India's most sacred rivers, Ma Ganga and Ma Yamuna. Here we witnessed the brittleness (i.e., lack of resiliency) of the sacred temple complexes to which pilgrims dramaturgically journey (aided largely by petroculture in so doing) and were introduced to a likely future where the origin point of the goddess worshipped at Gangotri may no longer exist once the glacier dries up. Same with our journey to Mecca in Saudi Arabia and a very likely future of 150 °F days. This heat will cripple the region and stress the power grid, water resources, and the ability to travel. How much harder to travel for the millions of pilgrims on *hajj* each year as they dramaturgically perform with their bodies one of the most central rituals of Islam?

After this we returned to the Himalayas and its water-stressed future, but this time to Ladakh. Here we encountered ecosystem peoples, omnivores, and ecological refugees, all, and Buddhist-inspired ice stupas as an effort to adapt to a changing water regime. We also met Buddhist monks advocating for a sustainable development that must rise to the challenges facing the region due to climate change.

After Ladakh we entered into post-materialist, posthuman Western spaces of religious dramaturgy inspired by and inspired to engage in prosustainable behaviors. This included visiting the oceans where surfers engage in aquatic nature religion. For some surfing can be a quest-religion that includes posthuman experiences of a sacred nature. For these surfers such experience may birth a responsibility inspired by the love of experiencing the ocean's agency to then dramaturgically act to protect it. From surfing we entered into the erotic posthuman assemblages of European-descendant ecosex practitioners in their post-materialist spaces. In these Western, posthuman erotic spaces we explored the potential of dramaturgical religious practices informing and contributing to prosustainable behaviors that may aid, if at all, in cultivating biocultural resilience as our home continues to warm. For ecosexuals we found a becoming-with response to petroculture based on queering bodies and thus a queering of petroculture that signals a move towards posthuman understandings and assemblages tethered to exhortations to preserve wild places and to consume less.

In terms of bodily movements and dramaturgies more broadly, the above can be seen as responses by and from bodies within "religious traditions as an enabling condition of meaning"—and that for the latter case studies suggest that "one's own movement making feeds a willingness and ability to move with the earth in mutually life-enabling ways" (LaMothe 2015, 206). This is because the dramaturgical bodily movements of those at Standing Rock, in Ladakh, in surfing, and other ecologies of place are enabling new conditions of meaning: posthuman conditions that with sincerity take intra-active agency and performative communication seriously, where even the land itself is part of such material co-becoming.

If such responses to create new religious modes of meaning and hoped for shifts to prosustainable behaviors will be enough to reverse the speeding up evolutionary unfolding of the Anthropocene remains to be seen. Trends suggest though that it is probably safest to assume these dramaturgical religious responses will be more and more geared towards deeply adapting to a future that every new climate study suggests will be that much harsher for our types of animal bodies. This sobering conclusion is the deep anchor for the questions asked, arguments presented, and scholarship undertaken throughout this book.

NOTE

1. Weise, Elizabeth. September 23, 2019. "'How dare you?' Read Greta Thunberg's emotional climate change speech to UN and world leaders." *USA Today*. https://www.usatoday.com/story/news/2019/09/23/greta-thunberg-tells-un-summit-youth-not-forgive-climate-inaction/2421335001/ Accessed July 18, 2020.

Afterword

We made it. I think. I hope? I strategically ponder with a "?" as what does it mean to "make it," when as I write this Afterword our planet is spiking with methane due to fracking, landfills, and agriculture. Methane that has a 20 times more powerful warming effect than CO2. Methane that could have entirely been avoided if we just switched our eating habits and did not invest in fracking technologies. What does it mean when in 2018 parts of Oman did not get below 105 °F for 51 hours straight? Or that our oceans are rapidly becoming warmer and warmer, saturated with heat, with half of that heat building over the last 20 years? Or that as a species we release the equivalent of 500,000 first-generation atomic bombs worth of greenhouse gas heat-trapping potential every day? And what does it mean when Siberia is spiking over 100 °F in July 2020 (even hotter than Dallas or Houston, Texas during the same time), moving us closer to a potential permafrost carbon (and methane) bomb of gigatons? When asked about these same record temperatures, the melting and thawing permafrost and ice, and the raging wildfires brought by increased heat and drought to the region, the senior research scientist at the National Snow and Ice Data Center at the University of Colorado at Boulder Walt Meier said, "We always expected the Arctic to change faster than the rest of the globe. But I don't think anyone expected the changes to happen as fast as we are seeing them happen."[1]

This insight is the mantra of our times. Study after study, in most places around the world, and lead scientists have a similar, if not the same, response. The modeled predictions of climate change are happening faster, and often, shifting in ways worse than even predicted. Pick your metric—rising temperatures (2020 tied with 2016 for the warmest recorded years on record, with average ocean temperature reaching the third highest recorded), ocean acidification, droughts, changing rainfall patterns, slowing ocean currents, changing animal and insect migrations, drying out forests—faster and faster. And despite nonbinding promises to protect 30% of the planet,[2] or the quickly dropping cost of solar, or potential of small nuclear reactors to diversify energy provision, the reality is that we have, assuming business as usual CO_2 emissions, about 20–30 years before a temperature threshold is crossed where about 50 percent of the CO_2 currently captured by terrestrial biomes will no longer be, thus moving this CO_2 from a "sink" to a "source" and therefore driving even more warming.[3]

Given this, I return to the same frustrations I voiced about religious studies and theology, and academia more broadly, at the beginning of the book: do we need another study on some obtuse lacunae in religion, given our planetary home is rapidly tipping into a potential hell scape unfit for human habitation in many extant places 20 to 50 to 75 to 100 years out from now? My own confirmation bias points me towards the shorter end of the timeline, but what's another twenty years when the end game is the same, and we are precariously getting closer and closer to "locked in" at this point no matter what?

Stopping climate change was a task of human animals in the 1950s; mitigating it a task for human animals in the 1970s and 1980s; adapting to it a task for human animals from the 1990s onwards. And this adaptation has to occur in infrastructure that was built for a different planetary climate regime. This means higher education needs to shift its modus operandi toward no longer producing consumers for a neoliberal industrial economy based on violence against the earth and bodies of color, but rather producing earth citizens capable of materially dwelling in ecologies of place and adapting to a maladaptive planetary regime shift. And do so on brittle campuses in bio-ecological places undergoing climate shifts. Do we have it in ourselves, and do the institutions within which we labor (here I include academic presses), to shift toward an embodied education of adaptive flourishing and resilient thriving in the chaos to come? Can we create an educational model that contributes to a bourgeoning (for non-indigenous peoples) traditional ecological knowledge attuned to right livelihood in bio-ecologies of place?

After exploring the animating questions and visiting the ecologies of place that inform this book, and while grappling with the above questions, I am left with thinking about the role of the scholar on a planet saturated with the poisonous (literal and figurative) material residues of petroculture. And I come back to an animating question of this book I pose to academia: How is this possible? How do we have thousands of, in terms of human history, over-educated human animals operating in ways that still seem to think we are separate from bio-ecological places? Or who apparently seem to think, given what gets published, and contrary to what this book explored, that what happens to micro- and meta-level biogeochemical systems and meso-level biocultural ecologies of place have nothing to do with entire disciplines?

Outside of still lamentably marginalized scholars of religion and nature, or religion and animals (both still seen as peripheral hires in departments, if at all), or ecotheology, how can the majority of two whole fields think that what has been published and researched in the past and continues to do in the present has anything to do with a planet without ice caps, passing multiple planetary boundaries, and where humans will be in massive ecological refugee flux? I ask these questions because all of us need to realize this: "As the threat to humanity of a powerful climatic *global* force of our own making has never before existed, much of what has been written in the past [and present] about [religion] is irrelevant to the future we now face" (i.e., this is my rewording of the quote shared in the book's Introduction from Glenn Albrecht).

We can easily substitute most any academic enterprise or discipline in for religion in this quote. Although the specific examples I lament and critiqued in the Introduction and here may shift down some in other fields, the same questions remain. This is because to date, academia as a whole, despite peripheral efforts from the Association for the Advancement of Sustainability in Higher Education,[4] the Sustainability Curriculum Consortium, and also Second Nature, is still not taking the material embeddedness of our praxis seriously. It is not taking the violence caused by the modern economy seriously, nor the politics of capitulation to this economy. If we did, I think we would collectively be much more concerned, and our collective scholarship and teaching would reflect the brittleness of the moment: that we are tipping into maladaptive climate regimes that will require a redesign of everything, including education and thus professional training, duties, and goals. In short, the academy does not need another obscure religious studies treatise on some random historical moment or figure that almost no one will ever read. This is not an indictment of just religious studies. It's an indictment of the whole system that rewards such efforts as the world literally burns. And again, to be clear and as shared in the Introduction—this is not at all an indictment of the intrinsic value of those laboring in the academy, or in religious studies and in theology, as committed scholars trying to navigate the dictates of wage-based capitalism by publishing or perishing.

It is a statement that we have bought into the ontological abstraction that there's a thing called "the academy," or "college," that actually exists, and it does so within something called "the economy." And that within this we are scholars and intellectuals, "trained" in a subdiscipline that we take to be materially real, all while somehow being protected from bio-ecological shifts that are going to severely impact us as actual evolved animals. This belief is held by scientists and humanists, both, where "from a neo-materialist perspective, [they both] have not infrequently worshipped at the very same altar: that of a disembodied mind and intellect that is largely abstract in its essence, whether it emerges from processing 'information' or linguistic 'discourse'" (LeCain 2017, 37).

It's like believing a corporation exists. Show me on the periodic table of elements where Harvard exists, or Google, or tenure, or the American Academy of Religion for that matter. They do not, by bio-ecological definition and as shaped by laws of thermodynamics. Rather, bodies exist, in material sympoeitic becoming, shaped by evolutionary scaffolding. Human bodies and biocultures exist in this space. My body. Your body. Those of your colleagues and students. In specific bio-ecologies with meso-level biocultural material shaping, where for some this shaping includes the performance of what we call the academy. So what type of biocultures do we want? What is the place and role of education in that, where this is more attuned to such bio-ecological realities and can therefore help us create adaptive resiliency? And how do we rethink of the performance of higher education while not remaining wedded to "benign strains of climate humanism [that] exhibit a striking dearth of feeling for a *living* Earth" (Sideris 2020, 12, italics original)?

Instead most of us think of education from within the context where for a few hundred years some humans within White CisHetero Settler Colonial contexts said there are things called "disciplines" and we study them and their claimed domains of data in buildings we build at places called universities. It may be that we need to reflect on what it means if this social construct does not have any biocultural adaptive value in the long-term view of planetary unfolding, especially when many of these built places will be underwater in the coming decades! Tenure will not mean much at that point, nor will another book on some random aspect of religion. The merits of even this book are but a miniscule drop in the bucket given this. This should not deter efforts—mine, my editor's, the reader's, our student's—to somehow try to change the above biocultural system. Such effort is entirely salutary and required for the engaged liberated scholarship in this book I have claimed is needed.

So we need a praxis—one built from within the academy, but that can also draw insights and paths of practice from some peoples outside the academy but for whom the below is already their long-standing practice (while of course being leery of appropriation). It is a praxis to which we can all contribute with our respective scholarship, teaching, and service. One fit for the liberation needed. One that takes feminist new materialist, queer, ecological animist, posthuman, and other analytical lenses used in this book to heart. One that begins with land, with the material bio-ecologies of place, and builds reciprocal relationships and knowledge systems with the agents and agencies of that place, guided by ceremony (Kimmerer 2013). A praxis of dwelling, akin to the dwelling perspective shared earlier in this book by Tim Ingold, moving us towards an ontology that understands the "universe [to be] a hundred percent relational" (Viveiros de Castro 2015, 255), within which humans are but one of many relational agental forces. Yet a praxis based in part upon "bad environmentalism," that recognizes imperfection, contradiction, the absurd and ambiguous, that moves beyond sentimentality and romanticization and guilt and shaming (Seymour 2018). And it must be a praxis that recognizes different cultures have different conceptualizations of nature (and agency), and all are needed to create effective environmental governance at global scales to help protect as much as we can all planetary bio-ecologies (Coscieme et al. 2020) and local ecologies of place. It needs to be a praxis geared towards an ecopoiesis—a "becoming of belonging" where the starting point is our habitats: "the web of constitutive interrelationships from which assemblages—communities, regions, institutions, concepts, families, individuals—emerge as singularities and which these assemblages, in turn, participate in composing" (Miller 2019, 132). These assemblages are biomaterial, influence and are part of biocultures, and contain agency in forming and shaping places of dwelling.

When thinking of assemblages, we must also think about performance, and how both pertain to a praxis of fidelity to ecologies of place. Those working, dwelling, and performing in academia must shift its focus to such fidelity of the bio-ecological places within which we labor. As Judith Butler explains, "if politics is oriented toward the making and preserving of the conditions that allow for livability, then it seems that the space of appearance is not ever fully separable from questions of infrastructure and architectures, and that they not only condition the action, but take part in the making of the space of politics" (2015, 127). The academic aspect of our praxis as paid educational professionals needs to see our workplaces as an architecture of new biopolitics, where posthuman becoming and insights are woven into efforts to adapt to runaway global warming.

Just as those practicing religion are shifting dramaturgical actions, as we saw in chapters 6 and 7, similar shifts must attend to the academy and the performance of our constructed disciplines. Currently those scholars willing to understand that to "discuss the posthuman is also to stare into the abyss of the inhumanity of our times" (Braidotti 2017, 84) are seemingly in the minority. Yet stare into this abyss we must, and in works on intersectional oppressions, already do. We must extend our gaze even further outwards, though, and realize that the hegemonic oil lifeways discussed in chapter 4 bring a violence to all bodies, including bodies of fellow-creature kinds yet born. We need a praxis of liberation for all these bodies, built upon a sympoietic Symbiocene: "a period in the history of humanity . . . characterized by human intelligence and praxis that replicate the symbiotic and mutually reinforcing life-reproducing forms and processes found in living systems" (Albrecht 2019, 102).

There is a possible pathway towards what a redesign of higher education for the Symbiocene will require as we journey deeper into a climate-changing Anthropocene. It comes from the leading antiracist scholar, Dr. Ibram Kendi who writes in his powerful *How to Be an Antiracist*, "I had to forsake the suasionist bred into me, of researching and educating for the sake of changing minds. I had to start researching and educating to change policy. The former strategy produces a public scholar. The latter produces public scholarship" (2019, 231). In this book I have employed both tactics he flags. For the suasionist tactic I have modeled what education via religion and nature/ecology research about bodies in a time of planetary regime shifts and changing bio-ecologies of place should resemble moving forward. Such modeling I argue applies to the entirety of the field of religious studies, and to theology, as well. There is no religion absent the material places where humans make and perform it, and those places are rapidly shifting to something hotter, less capable of letting human lifeways continue. Such religious studies education that informs the public about the dangers of climate change to our livelihoods can aid the larger academy in helping to change minds about why we need to put all our educational efforts towards climate justice and adaptation.[5] In short, we must all become suasionist public scholars on climate warming; and the intersectional impacts such warming is already and will continue to have on minoritized bodies, as well as on bio-ecological places. This need is for all disciplines in academia and must be supported by administrators throughout the academy, as well as by funding agencies.

If a course is taught that does not address melting permafrost, collapsing ice sheets, the impact of climate change on the global economy, the need to redesign how we get and use energy and food, and ongoing violence on certain "otherized" bodies, then it is time to question the value of such a course for the approximately 300,000 years old biocultural animals living in bio-ecological habitats that we always have been. If a research project neglects to factor in climate change, or a publication does not try to help us think of pathways towards adaptive resilience, then it is time to question the value of such research for the same animals and our to-date short 300,000 year geological tenure on earth.

Here again I return to my initial framing of the book—in this my criticism is not with colleagues throughout academia. It is with the structural and policy demands and impediments we labor within, and the inherited history of Western academe that has never, and still largely does not, take bio-ecological places seriously. It is with academic policies and practices that are thoroughly anthropocentric, that perpetuate the insidious ills of White CisHetero Settler Colonialism, and that operate as if "disciplines" are real things that we can touch and feel and that are born and die. In short, our understanding of public scholarship is wedded to a view of education that is at odds with the reality of our times, and thus our public scholarship needs to "adapt or die," to paraphrase a popular bumper sticker from the 1970s. We need faculty, chairs, Deans, Provosts, Presidents, and funders who can lead on such insights and help rapidly redesign education for "after nature" (Purdy 2015). And if these people are currently in short supply—which they are—then we need to become them.

Which brings me to the second of Kendi's insights. We need to start researching and educating to change policy. Whether that policy is economic (e.g., advocating for a carbon tax), educational, entrepreneurial, based on intersectional identities, or environmental, we need to be involved with researching and educating on it. Our research and education should be geared towards empowering a shift in policies in all policy domains toward adaptive resilience, equity, and justice. This shifts our role from just being public scholars, to producing public scholarship. Given this role, it is hoped that this book will have some type of impact on changing the policies of education so they become more attuned with the biogeochemical reality of our Anthropocene times. Even if it is one department of religion somewhere creating a department level student learning outcome that says, "Majors will become advocates for policy-based adaptive climate solutions through a religion and nature/ecology lens," then it will have done this part of its job. However I hope that the public scholarship aspect of this book joins with a growing chorus of colleagues throughout the academy advocating for sane climate policies and solutions, so a biocultural tipping point can be rapidly reached that helps direct higher education towards supporting and enacting sustainable dwelling and resilient transformations in the ecologies of place where we perform our education. And if the game is lost as we are 50 years too late, then public scholars and their public scholarship need to point towards graceful extinction (Bird Rose et al. 2017; van Dooren 2014) as petroculture collapses, and ways to deal with the continued violence and trauma that comes with petroculture and its end game.

NOTES

1. "Rapid Arctic meltdown in Siberia alarms scientists." *The Washington Post.* Isabelle Khurshudyan, Andrew Freeman, and Brady Dennis. July 3, 2020. Accessed July 4, 2020. https://www.washingtonpost.com/climate-environment/rapid-arctic-m eltdown-in-siberia-alarms-scientists/2020/07/03/4c1bd6a6-bbaa-11ea-bdaf-a129f921 026f_story.html?fbclid=IwAR0HkeN-wz_9gJly_Eq4kZNduGjOWsRGrhBzG6NX RVOBFs27XaPT2tQLc-A

2. "50 countries vow to protect 30% of land and sea by 2030." *AP News.* Sylvie Corbet. January 11, 2021. Accessed January 19, 2021. https://apnews.com/articl e/50-countries-protect-land-sea-climate-3450a3ef2bafb1149b7bb0931914eefc#:~:te

xt=PARIS%20(AP)%20%E2%80%94%20At%20least,at%20protecting%20the%20 world's%20biodiversity.

3. "How Close are we to the Temperature Tipping Point of the Terrestrial Biosphere?" Katharyn Duffy, et al. *Science Advances:* 7.3. January 13, 2021. Accessed January 19, 2021. https://advances.sciencemag.org/content/7/3/eaay1052

4. For efforts taken by AAR, see Bobbi Patterson and Robert Puckett (2015).

5. In the United States this goal will be hampered in Republican-controlled legislatures, as according to a 2019 Pew Research 59 percent of Republicans surveyed think college/higher education is now bad for Americans. The strong correlation with Republicans denying human-induced climate change should not be forgotten when thinking about the implications of the above data point. See Robby Soave, "Survey: 59% of Republicans Now Think College is Bad for America." August 19, 2019. Accessed July 4, 2020 https://reason.com/2019/08/19/pew-survey-republicans-co llege-campus-safe-spaces/. In short, it behooves all of us to see that ideology on both education and science as held by the majority of currently polled Republicans has become maladaptive to species survival, and to overall planetary health.

Coda

I close here by citing one of my heroes, whose writing informs who I am as a public scholar, and thus informs my public scholarship. It comes from the ecological agrarian, poet, and essayist, Wendell Berry, who in 1979 wrote to his friend the bioregional philosopher Gary Snyder:

> It is apparently now possible to learn how to join one's work *properly* to the universe. Symbiosis becomes the ideal or standard, which applies to any kind of human product—a house, a factory, a farm, a poem. And if one undertakes to join one's work properly to the universe, then one has immediately at hand a critical standard. Abusable, like any standard—for some jerk is bound to assume that what's wanted is symbiotic *propaganda*. The corrective, I assume, would be a propriety of another sort—a fit sense of the complexity of the universe and its relationships, so great as to have to be acknowledged ultimately as mystery. (2014, 40, italics in original)

We return, in whatever time is left to us, to the ground that never forgot us: to our bio-ecological places we (nonindigenous dwellers) have abstracted ourselves from to our own detriment. We return to our material, animal bodies that dwell in agental mystery in ecologies of place. In this mystery, we ask what is a standard of propriety that can be used to investigate the value of our current system of higher education? How can we learn how to join the work of the academy, and within that, religious studies (and theology), properly to the universe? Hopefully this book has provided one possible pathway to the work of a performed academic religious studies scholarship worthy of a standard and praxis of symbiosis: a fusing of human liberation with planetary health, offered in service of mystery, resilience, adaptation, and healing, attuned to the suffering already past, currently ongoing, and that will only become worse with the catastrophic warming to come.

Appendix

Table 1   Emerging Academic Views of Bodies, adapted from Francesca Ferrando (2013)

Movement type beyond the inherited view of an Anthropocentric, Isolated Humanistic Body:	Brief Definition/Summary of:	Selected branches of:
Transhumanism	Science and technology can be used to enhance biological human bodies-medicine, nanotechnology, life extension, mind uploading, cryonics	Libertarian, democratic, extropian
Posthumanism (used throughout this book)	Removing strict boundaries between human/nonhuman animals, biological organisms, and even machines, and between physical and nonphysical realms; effort to do away with any form of primacies, exceptionalisms, or dualisms in understanding of bodies	A praxis of ethics and applied philosophy to redefine ontology of "self"; branches include feminist, literary, critical, philosophical, technological
New Materialisms (used throughout this book)	Matter is re-inscribed as shaping and constituting agency; all biology (and thus culture, and thus language) is materialistically constructed via ongoing processes	Feminist, literary, philosophical, political, critical

(Continued)

Table 1 (Continued)

Queer Ecologies (used throughout this book)	Movement beyond binaries and viewing organisms as distinct, and species as bounded; critique of heterosexist normative assumptions when theorizing and studying bio-ecologies and the organisms within
Antihumanism	Deconstruction of the notion of the human (and the latter's construction of abiding faith in progress)
Metahumanism	Body is locus for amorphic re-significations occurring in kinetic relations as body-network (ibid: 32)
Posthumanities	In academia, "an internal shift extending the study of the human condition to the posthuman; it may also refer to future generations of beings evolutionarily related to the human species" (ibid: 32)

Bibliography

Aanen, Duur, and Paul Eggleton. 2017. "Symbiogenesis: Beyond the Endosymbiosis Theory?" *Journal of Theoretical Biology* 434 (7): 99–103.

Adams, Carol, and Josephine Donovan. 2007. "Introduction." In *The Feminist Care Tradition in Animal Ethics*, edited by Carol Adams and Josephine Donovan, 1–20. New York: Columbia University Press.

Agard-Jones, Vanessa. 2013. "Bodies in the System." *Small Axe* 42: 182–192.

Akimichi, Tomoya. 1996. "Image and Reality at Sea: Fish and Cognitive Mapping in Carolinean Navigational Knowledge." In *Redefining Nature: Ecology, Culture and Domestication*, edited by Roy Ellen and Katsuyoshi Fukui, 493–514. Washington, D.C.

Alaimo, Stacy. 2016. *Exposed: Environmental Politics & Pleasures in Posthuman Times*. Minneapolis: University of Minneapolis Press.

Alaimo, Stacy, and Susan Hekman. 2008. "Introduction: Emerging Models of Materiality in Feminist Theory." In *Material Feminisms*, edited by Stacy Alaimo and Susan Hekman, 1–19. Bloomington: Indiana University Press.

Albanese, Catherine. 1991. *Nature Religion in America: From the Algonkian Indians to the New Age*. Chicago: University of Chicago Press.

Albrecht, Glen. 2005. "Solastalgia: A New Concept in Human Health and Identity." *PAN (Philosophy, Activism, Nature)* 3: 44–59.

Alley, Richard. 2011. *Earth: The Operator's Manual*. New York: W.W. Norton & Company.

Almazroui, Mansour. 2013. "Simulation of Present and Future Climate of Saudi Arabia Using a Regional Climate Model (PRECIS)." *International Journal of Climatology* 33: 2247–2259.

Almazroui, Mansour, M. Nazrul Islam, H. Athar, P. D. Jones, and M. Ashfaqur Rahman. 2012. "Recent Climate Change in the Arabian Peninsula: Annual Rainfall and Temperature Analysis of Saudi Arabia for 1978–2009." *International Journal of Climatology* 32: 953–966.

AlSarmi, Said, and Richard Washington. 2011. "Recent Observed Climate Change Over the Arabian Peninsula." *Journal of Geophysical Research* 116: 1–15.

Anderson, P. W. 1972. "More is Different." *Science* 177 (4047): 393–396.

Armstrong, Karen. 2015. *Fields of Blood: Religion and the History of Violence*. New York: Anchor Books.

Asberg, Cecilia, Kathrin Thiele, and Iris Van Der Tuin. 2015. "Speculative *Before* the Turn: Reintroducing Feminist Materialist Performativity." *Cultural Studies Review* 21 (2): 145–172.

Atran, Scott. 2002. *In Gods We Trust: The Evolutionary Landscape of Religion*. New York: Oxford University Press.

Banerjee-Guha, Swapna. 2010. "Introduction: Transformative Cities in the New Global Order." In *Accumulation by Dispossession: Transformative Cities in the New Global Order*, edited by Swapna Banerjee-Guha, 1–16. New Delhi: SAGE.

Barad, Karen. 2008. "Posthumanist Performativity." In *Material Feminisms*, edited by Stacy Alaimo and Susan Hekman, 120–154. Bloomington: Indiana University Press.

Basso, Keith. 1996 *Wisdom Sits in Places: Landscape and Language Among the Western Apache*. Albuquerque, NM: University of New Mexico Press.

Baugh, Amanda. 2017. *God and the Green Divide: Religious Environmentalism in Black and White*. Berkeley: University of California Press.

Bauman, Whitney. 2014. *Religion and Ecology: Developing a Planetary Ethic*. New York: Columbia University Press.

Bauman, Whitney. 2016. "What's Left (Out) of the Lynn White Narrative?" In *Religion and Ecological Crisis: The "Lynn White Thesis" at 50*, edited by Todd LeVasseur and Anna Peterson, 165–177. New York: Routledge.

Bauman, Whitney. 2017. "The Ethics of Wicked Problems: Entanglement, Multiple Causality and Rainbow Time." *Worldviews: Environment, Culture, Religion* 21: 7–20.

Bauman, Whitney. 2018. "Queer Values for a Queer Climate: Developing a Versatile Planetary Ethic." In *Meaningful Flesh: Reflections on Religion and Nature for a Queer Planet*, edited by Whitney Bauman, 103–123. Earth: Punctum Books.

Bauman, Whitney, and Lisa Stenmark. 2018. "Introduction." In *Unsettling Science and Religion: Contributions and Questions from Queer Studies*, edited by Whitney Bauman and Lisa Stenmark, 1–31. New York: Lexington Books.

Baviskar, Amita. 2005. *In the Belly of the River: Tribal Conflicts over Development in the Narmada Valley*. New York: Oxford University Press.

Bekoff, Marc. 2006. "Wild Justice, Social Contagion, Fairness, and Morality: A Deep Appreciation for the Subjective Lives of Animals." In *A Communion of Subjects: Animals in Religion, Science & Ethics*, edited by Paul Waldau and Kimberley Patton, 461–480. New York: Columbia University Press.

Bell, Catherine. 1998. "Performance." In *Critical Terms for Religious Studies*, edited by Mark Taylor, 205–224. Chicago: The University of Chicago Press.

Bellah, Robert. 2011. *Religion in Human Evolution: From the Paleolithic to the Axial Age*. Cambridge, MA: Harvard University Press.

Bender, Barbara. 2002. "Landscape and Politics." In *The Material Culture Reader*, edited by Victor Buchli, 135–174. Oxford, UK: Berg.

Bennett, Jane. 2010a. "A Vitalist Stopover on the Way to a New Materialism." In *New Materialisms: Ontology, Agency, and Politics*, edited by Diana Coole and Samantha Frost, 47–69. Durham: Duke University Press.

Bennett, Jane. 2010b. *Vibrant Matter: A Political Ecology of Things*. Durham: Duke University Press.

Berkes, Fikret. 1999. *Sacred Ecology: Traditional Ecological Knowledge and Resource Management*. Philadelphia: Taylor and Francis.

Berry, Wendell. 1977. *The Unsettling of America: Culture & Agriculture*. San Francisco: Counterpoint.

Berry, Wendell, and Gary Snyder. 2015. *Distant Neighbors: The Selected Letters of Wendell Berry and Gary Snyder*. Edited by Chad Wriglesworth. Berkeley: Counterpoint.

Bird-David, Nurit. 2000. "'Animism' Revisited: Personhood, Environment, and Relational Epistemology." In *Indigenous Religions: A Companion*, edited by Graham Harvey, 72–100. New York: Bloomsbury.

Bird Rose, Deborah, Thom van Dooren, and Matthew Chrulew. 2017. *Extinction Studies: Stories of Time, Death, and Generations*. New York: Columbia University Press.

Bonnedahl, Karl, and Pasi Heikkurinen, eds. 2019. *Strongly Sustainable Societies: Organising Human Activities on a Hot and Full Earth*. New York: Earthscan.

Borne, Gregory, and Jess Ponting. 2015. *Sustainable Stoke: Transitions to Sustainability in the Surfing World*. Plymouth: University of Plymouth Press.

Boudinot, Garrett, and Todd LeVasseur. 2016. "'Grow the Scorched Ground Green.' Values and Ethics in the Transition Movement." *Journal for the Study of Religion, Nature and Culture* 10 (3): 376–401.

Boyer, Pascal. 2007. *Religion Explained: The Evolutionary Origins of Religious Thought*. New York: Basic Books.

Bradshaw, Corey, P. R. Ehrlich, A. Beattie, G. Ceballos, E. Crist, J. Diamond, R. Dirzo, A. H. Ehrlich, J. Harte, M. E. Harte, and G. Pyke. 2021. "Underestimating the Challenges of Avoiding a Ghastly Future." *Frontiers in Conservation Science* 1 (615419): 1–10. DOI: 10.3389/fcosc.2020.615419.

Braidotti, Rosi. 2017. "Critical Posthuman Knowledges." *The South Atlantic Quarterly* 116 (1): 83–96.

Braidotti, Rosi, and Cosetta Veronese. 2017. "Can the Humanities Become Posthuman? A Conversation." In *Environmental Humanities: Voices from the Anthropocene*, edited by Serpil Oppermann and Serenella Iovino, 339–346. New York: Rowman & Littlefield.

Braungart, Michael, and William McDonough. 2002. *Cradle to Cradle: Remaking the Way We Make Things*. New York: North Point Press.

Broome, John. 2012. *Climate Matters: Ethics in a Warming World*. New York: W. W. Norton & Company.

Brown, Adrienne Maree. 2019. *Pleasure Activism: The Politics of Feeling Good*. Edinburgh: AK Press.

Brown, Patrick, and Ken Caldeira. 2017. "Greater Future Global Warming Inferred from Earth's Recent Energy Budget." *Nature* 552 (December 7): 45–64.

Butcher, Andrea. 2013a. "Tulku and Deity Assistants." *International Association for Ladakh Studies* 30: 16–24.

Butcher, Andrea. 2013b. "Keeping the Faith: Divine Protection and Flood Prevention in Modern Buddhist Ladakh." *Worldviews: Global Religions, Culture, and Ecology* 17: 103–114.

Butler, Judith. 1988. "Performative Acts and Gender Constitution: An Essay in Phenomenology and Feminist Theory." *Theatre Journal* 40 (4): 519–531.

Butler, Judith. 1990. *Gender Trouble: Feminism and the Subversion of Identity*. New York: Routledge.

Byg, Anja, and Jan Salick. 2009. "Local Perspectives on a Global Phenomenon—Climate Change in Eastern Tibetan Villages." *Global Environmental Change* 19: 156–166.

Candlin, Fiona, and Raiford Guins. 2009. *The Object Reader*. London: Routledge.

Capra, Fritjof, and Pier Luigi Luisi. 2014. *The Systems View of Life: A Unifying Vision*. New York: Cambridge University Press.

Carp, Richard. 2012. "Editor's Introduction." In *Dancing Culture Religion*, edited by Sam Gill, front page. Lanham: Lexington Books.

Carp, Richard. 2014. "Material Culture." In *The Routledge Handbook of Research Methods in the Study of Religion*, edited by Michael Stausberg and Steven Engler, 474–490. New York: Routledge.

Casey, Edward. 1997. *The Fate of Place: A Philosophical History*. Berkeley: University of California Press.

Castelli, Elizabeth. 2001. "Introduction." In *Women, Gender, Religion: A Reader*, edited by Elizabeth Castelli, 3–25. New York: Palgrave.

Cela-Conde, Camilo, and Francisco Ayala. 2007. *Human Evolution: Trails from the Past*. New York: Oxford University Press.

Chidester, David, and Edward Linenthal, eds. 1995. *American Sacred Space*. Bloomington: Indiana University Press.

Chowdhury, Shakhawat, and Muhammad Al-Zahrani. 2013. "Implications of Climate Change on Water Resources in Saudi Arabia." *Arabian Journal of Science and Engineering* 38: 1959–1971.

Clare, Stephanie. 2019. *Earthly Encounters: Sensation, Feminist Theory, and the Anthropocene*. Albany: SUNY Press.

Clark, John. 2011. *Hawaiian Surfing: Traditions from the Past*. Honolulu: University of Hawai'i Press.

Clouse, Carey. 2016. "Frozen Landscapes: Climate-Adaptive Design Interventions in Ladakh and Zanskar." *Landscape Research* 41 (8): 821–837.

Clouse, Carey. 2017. "The Himalayan Ice Stupa: Ladakh's Climate-adaptive Water Cache." *Journal of Architectural Education* 71 (2): 247–251.

Coleman, Simon, and John Eade. 2004. "Introduction: Reframing Pilgrimage." In *Reframing Pilgrimage: Cultures in Motion*, edited by Simon Coleman and John Eade, 1–25. New York: Routledge.

Comer, Krista. 2010. *Surfer Girls in the New World Order*. Durham: Duke University Press.

Connelly, William. 2017. *Facing the Planetary: Entangled Humanism and the Politics of Swarming*. Durham: Duke University Press.

Cronon, William. 1996. "The Trouble with Wilderness: Or, Getting Back to the Wrong Nature." *Environmental History* 1 (1): 7–28.

Dawson, Kevin. 2018. *Undercurrents of Power: Aquatic Culture in the African Diaspora*. Philadelphia: University of Pennsylvania Press.

Deegan, Chris. 2000. "The Narmada: Circumambulation of a Sacred Landscape." In *Hinduism and Ecology: The Intersection of Earth, Sky, and Water*, edited by Christopher Key Chapple and Mary Evelyn Tucker, 389–399. Cambridge: Harvard University Press.

Deloria, Vine, Jr. 1994. *God is Red: A Native View of Religion*. Golden: Fulcrum Publishing.

DeNicola, Erica, Omar S. Aburizaiza, Azhar Siddique, Haider Khwaja, and David O. Carpenter. 2015. "Climate Change and Water Scarcity: The Case of Saudi Arabia." *Annals of Global Health* 81 (3): 342–353.

Depledge, Joanna. 2008. "Striving for No: Saudi Arabia in the Climate Change Regime." *Global Environmental Politics* 8 (4): 9–35.

De Vos, Jurriaan, Lucas N. Joppa, John L. Gittleman, Patrick R. Stephens, and Stuart L. Pimm. 2014. "Estimating the Normal Background Rate of Species Extinction." *Conservation Biology* 20, 2 (April): 452–462.

de Waal, Frans. 2016. Stephen Macedo and Josiah Ober, eds. *Primats and Philosophers; How Morality Evolved*. Princeton: Princeton Science Library.

Diamond, Jared. 1997. *Guns, Germs, and Steel: The Fates of Human Societies*. New York: W.W. Norton.

Drew, Georgina. 2012. "A Retreating Goddess? Conflicting Perceptions of Ecological Change near the Gangotri-Gaumukh Glacier." *Journal for the Study of Religion, Nature and Culture* 6 (3): 344–362.

Dunbar, Robin. 2016. *Human Evolution: Our Brains and Behavior*. New York: Oxford University Press.

Eck, Diana. 1993. *Darśan: Seeing the Divine Image in India* (3rd Edition). New York: Columbia University Press.

Eck, Diana. 1996. "Ganga: The Goddess Ganges in Hindu Sacred Geography." In *Devi: Goddesses of India*, edited by John Stratton Hawley and Donna Marie Wulff, 137–154. Berkeley: University of California Press.

Edwards, Andres. 2010. *Thriving Beyond Sustainability: Pathways to a Resilient Society*. Gabriola Island: New Society Publishers.

Ehrlich, Paul. 2000. *Human Natures: Genes, Cultures, and the Human Prospect*. New York: Penguin Books.

Epstein, Barbara. 1997. "Postmodernism and the Left." *New Politics* 6 (2).

Evers, Clifton. 2009. "'The Point': Surfing, Geography and a Sensual Life of Men and Masculinity on the Gold Coast, Australia." *Social & Cultural Geography* 10 (8): 893–908.

Ferrando, Francesca. 2013. "Posthumanism, Transhumanism, Antihumanism, Metahumanism, and New Materialisms." *Existenz: An International Journal in Philosophy, Religion, Politics, and the Arts* 8 (2): 26–32.

Figueres, Christiana, and Tom Rivett-Carnac. 2020. *The Future We Choose: Surviving the Climate Crisis.* New York: Alfred A. Knopf.

Finney, Carolyn. 2014. *Black Faces, White Spaces: Reimagining the Relationship of African Americans to the Great Outdoors.* Chapel Hill: The University of North Carolina Press.

Flynn, Pierce 1987. "Waves of Semiosis: Surfing's Iconic Progression." *The American Journal of Semiotics* 5 (3–4): 397–408, 411–418.

Food and Agriculture Organization of the United Nations. 2006. "Livestock's Long Shadow: Environmental Issues and Options." http://www.fao.org/docrep/010/a0701e/a0701e00.HTM.

Foucault, Michel, and Paul Rabinow. 1984. *The Foucault Reader.* New York: Pantheon Books.

Fuller, Jason. 2011. "The Accidental Pilgrim: Vaiṣṇava tīrthas and the Experience of the Sacred." In *Studying Hinduism in Practice*, edited by Hilary Rodrigues, 62–74. New York: Routledge.

Fuller, Robert. 2008. *Spirituality in the Flesh: Bodily Sources of Religious Experience.* New York: Oxford University Press.

Fuller, Robert. 2013. *The Body of Faith: A Biological History of Religion in America.* Chicago: The University of Chicago Press.

Gaard, Greta. 1997. "Toward a Queer Ecofeminism." *Hypatia* 12 (1): 137–156.

Gadgil, Madhav, and Ramachandra Guha. 1992. *This Fissured Land: An Ecological History of India.* Oxford, UK: Oxford University Press.

Ghosh, Amitav. 2015. *The Great Derangement: Climate Change and the Unthinkable.* India: Cyber City.

Globus-Veldman, Robin. 2019. *The Gospel of Climate Skepticism: Why Evangelical Christians Oppose Action on Climate Change.* Berkeley: University of California Press.

Globus-Veldman, Robin, Andrew Szasz, and Randolph Haluza-Delay. 2014. *How the World's Religions are Responding to Climate Change.* New York: Routledge.

Goodwin, Brian. 1994. *How the Leopard Changes Its Spots: The Evolution of Complexity.* New York: Charles Scribner's Sons.

Gould, Rebecca Kneale. 2005. *At Home in Nature: Modern Homesteading and Spiritual Practice in America.* Berkeley: University of California Press.

Grim, John, and Mary Evelyn Tucker. 2014. *Ecology and Religion.* Washington, D.C.: Island Press.

Grusin, Richard. 2015. "Introduction." In *The Nonhuman Turn*, edited by Richard Grusin, vii–xxix. Minneapolis: University of Minnesota Press.

Guerrero, M. A. Jaimes. 2000. "Native Womanism: Exemplars of Indigenism in Sacred Traditions of Kinship." In *Indigenous Religions: A Companion*, edited by Graham Harvey, 37–55. New York: Bloomsbury.

Guha, Ramachandra. 1990. *The Unquiet Woods: Ecological Change and Peasant Resistance in the Himalaya.* Berkeley: University of California Press.

Guha, Ramachandra. 2006. *How Much Should a Person Consume? Thinking Through the Environment*. Ranikhet: Permanent Black.

Gunderson, Lance, and C. S. Holling, eds. 2001. *Panarchy: Understanding Transformations in Human and Natural Systems*. Washington, D.C.: Island Press.

Gunderson, Lance, and Craig Allen. 2010. "Introduction: Why Resilience? Why Now?" In *Foundations of Ecological Resilience*, edited by Lance Gunderson, Craig Allen, and C. S. Holling, xiii–xxv. Washington, D.C.: Island Press.

Haberman, David. 2005. *River of Love in an Age of Pollution: The Yamuna River of Northern India*. Berkeley: University of California Press.

Haidt, Jonathan. 2012. *The Righteous Mind: Why Good People Are Divided by Politics and Religion*. New York: Vintage Books.

Hall, David, ed. 1997. *Lived Religion in America: Toward a History of Practice*. Princeton: Princeton University Press.

Halperin, Ehud. 2017. "Winds of Change: Religion and Climate in the Western Himalayas." *Journal of the American Academy of Religion* 85 (1): 64–111.

Haraway, Donna. 1996. *Simians, Cyborgs, and Women: The Reinvention of Nature*. London: Free Association Books.

Haraway, Donna. 2016. *Staying with the Trouble: Making Kin in the Chthulucene*. Durham: Duke University Press.

Harding, Stephan. 2006. *Animate Earth: Science, Intuition and Gaia*. White River Junction: Chelsea Green Publishing Company.

Harney, Stefano, and Fred Moten. 2013. *The Undercommons: Fugitive Planning & Black Study*. New York: Minor Compositions/Autonomedia.

Harris, Melanie Harris. 2017. "Introduction." In *Ecowomanism, Religion and Ecology*, edited by Melanie Harris, 4–12. Leiden: Brill.

Harrod, Howard. 2000. *The Animals Came Dancing: Native American Sacred Ecology and Animal Kinship*. Tucson: The University of Arizona Press.

Harvey, David. 1990. *The Condition of Postmodernity: An Enquiry into Cultural Change*. Oxford: Blackwell Publishing.

Harvey, David. 1996. *Justice, Nature & the Geography of Difference*. Cambridge: Blackwell.

Hawken, Paul, ed. 2017. *Drawdown: The Most Comprehensive Plan Ever Proposed to Reverse Global Warming*. New York: Penguin Books.

Heinberg, Richard. 2010. "What is Sustainability?" In *The Post Carbon Reader: Managing the 21st Century's Sustainable Crises*, edited by Richard Heinberg and Daniel Lerch. Healdsburg: Watershed Media.

Holling. C.S. 2001. "Understanding the Complexity of Economic, Ecological, and Social Systems." *Ecosystems* 4 (5): 390–405.

Hopkins, Rob. 2019. *From What is to What If: Unleashing the Power of the Imagination to Create the Future We Want*. White River Junction: Chelsea Green Publishing.

Hough-Snee, Dexter, and Alexander Eastman. 2017. *The Critical Surf Studies Reader*. Durham: Duke University Press.

Huber, Matthew. 2011. "Enforcing Scarcity: Oil, Violence, and the Making of the Market." *Annals of the Association of American Geographers* 101 (4): 816–826.

Huber, Toni. 1997. "Green Tibetans: A Brief Social History." In *Tibetan Culture in the Diaspora*, edited by Frank Korom, 103–119. Vieinna: Verlag.

Hultman, Martin, and Paul Pulé. 2018. *Ecological Masculinities: Theoretical Foundations and Practical Guidance.* New York: Routledge.

Humberstone, Barbara. 2011. "Embodiment and Social and Environmental Action in Nature-Based Sport: Spiritual Spaces." *Leisure Studies* 30 (4): 495–512.

Ingold, Tim. 2000. *The Perception of the Environment: Essays on Livelihood, Dwelling and Ski'l.* New York: Routledge.

Ingold, Tim. 2005. "Epilogue: Towards a Politics of Dwelling." *Conservation & Society* 3 (2): 501–508.

Ingold, Tim. 2011. *Being Alive: Essays on Movement, Knowledge and Description.* New York: Routledge.

Iovino, Serenella, and Serpil Oppermann. 2012. "Material Ecocriticism: Materiality, Agency, and Models of Narrativity." *Ecozon@* 3 (1): 75–91.

Isaac, Rhys. 1982/1999. *The Transformation of Virginia: 1740–1790.* Chapel Hill: University of North Carolina Press.

Ivakhiv, Adrian. 2003. "Orchestrating Sacred Space: Beyond the 'Social Construction' of Nature." *Ecotheology* 8 (1): 11–29.

Jackson, Michael. 2017. *How Lifeworlds Work: Emotionality, Sociality & the Ambiguity of Being.* Chicago: University of Chicago Press.

Jackson, Wes. 2010. *Consulting the Genius of the Place: An Ecological Approach to a New Agriculture.* Berkeley: Counterpoint.

Jackson, Wes. 2011. *Consulting the Genius of the Place: An Ecological Approach to a New Agriculture.* San Francisco: Counterpoint.

Jamieson, Dale. 2014. *Reason in a Dark Time; Why the Struggle Against Climate Change Failed—And What It Means for Our Future.* New York: Oxford University Press.

Jenkins, Willis. 2017. "Feasts of the Anthropocene: Beyond Climate Changes as Special Object in the Study of Religion." *The South Atlantic Quarterly* 116 (1): 69–81.

Jenkins, Willis, Even Berry, and Luke Beck Kreider. 2018. "Religion and Climate Change." *Annual Review of Environment and Resources* 43: 85–108.

Johnson, Linda. 1994. *Daughters of the Goddess: The Women Saints of India.* St. Paul, Minnesota: Yes International Publishers.

Johnston, Lucas. 2013. *Religion and Sustainability: Social Movements and the Politics of the Environment.* Bristol: Equinox Publishing.

Johnston, Lucas. 2016. "Cultivating an Academy We Can Live With: The Humanities and Education for Sustainability." *Religions* 7 (20): 1–16.

Justice, Christopher 1997. *Dying the Good Death: The Pilgrimage to Die in India's Holy City.* Albany: State University of New York Press.

Kendi, Ibram. 2019. *How to Be an Antiracist.* New York: One World.

Kershaw, Baz. 2012. "'This is the Way the World Ends, Not...?' On Performance Compulsion and Climate Change." *Performance Research* 17 (4): 5–17.

Klare, Michael. 2001. *Resource Wars: The New Landscape of Global Conflict.* New York: Metropolitan/Owl Book.

Kleiner, Catherine. 2003. "Nature's Lovers: The Erotics of Lesbian Land Communities in Oregon, 1974–1984." In *Seeing Nature Through Gender*, edited by Virginia Scharff, 242–262. Lawrence: University Press of Kansas.

Kohler, Timothy, Michael E. Smith, Amy Bogaard, Gary M. Feinman, Christian E. Peterson, Alleen Betzenhauser, Matthew Pailes, E. C. Stone, A. M. Prentiss, T. J. Dennehy, and L. J. Ellyson. 2017. "Greater Post-Neolithic Wealth Disparities in Eurasia than in North America and Mesoamerica." *Nature* 551 (November 30): 619–627.

Kohn, Eduardo. 2013. *How Forests Think: Toward an Anthropology Beyond the Human*. Berkeley: University of California Press.

Kubiak, Anthony. 2012. "Animism: Becoming-Performance, or Does This Text Speak to You?" *Performance Research* 17 (4): 52–60.

Lal, Vinay. 2015. "Climate Change: Insights from Hinduism." *Journal of the American Academy of Religion* 83 (2): 388–406.

LaMothe, Kimerer. 2015. *Why We Dance: A Philosophy of Bodily Becoming*. New York: Columbia University Press.

Lane, Belden. 2001. *Landscapes of the Sacred: Geography and Narrative in American Spirituality*. Baltimore: The Johns Hopkins University Press.

Lanza, Robert. 2010. *Biocentrism: How Life and Consciousness Are the Keys to Understanding the True Nature of the Universe*. Dallas: Benbella Books, Inc.

Latour, Bruno. 1993. *We Have Never Been Modern*. Cambridge: Harvard University Press.

Lawler, Kristin. 2010. *The American Surfer: Radical Culture and Capitalism*. New York: Routledge.

LeCain, Timothy. 2015. "Against the Anthropocene: A Neo-Materialist Perspective." *International Journal for History, Culture and Modernity* 3 (1): 1–28.

LeVasseur, Todd. 2014. "Environmental Philosophy in a Post-Ice Cap North Polar World." *Environmental Ethics* 36 (3): 303–318.

LeVasseur, Todd. 2015a. "Is Trash Hybrid?" *Green Humanities* 1: 75–103.

LeVasseur, Todd. 2015b. "'The Earth is *Sui Generis*': Destabilizing the Climate of Our Field." *Journal of the American Academy of Religion* 83 (2): 300–319.

LeVasseur, Todd. 2015c. "Introduction." *Journal of the American Academy of Religion* 83 (2): 297–319.

LeVasseur, Todd. 2017. *Religious Agrarianism and the Return of Place: From Values to Practice in Sustainable Agriculture*. Albany: SUNY Press.

LeVasseur, Todd. 2020. "Reverend Billy and the Church of Stop Shopping: Contemporary Religious Production on a Planet Passing Tipping Points." *Nova Religio: The Journal of Alternative and Emergent Religions* 23 (3): 86–109.

LeVasseur, Todd, and Bernard Zaleha. 2019. "When Christ is Maladaptive." In *Christian Theology and Climate Change*, edited by Ernst Conradie and Hilda Koster, 120–131. New York: T & T Clark/Bloomsbury.

Lewontin, Richard, and Richard Levins. 2007. *Biology Under the Influence: Dialectical Essays on Ecology, Agriculture, and Health*. New York: Monthly Review Press.

Lopez, Donald. 1998. *Prisoners of Shangri-La: Tibetan Buddhism and the West*. Chicago: University of Chicago Press.

Machon, Josephine. 2009. *(Syn)aesthetics: Redefining Visceral Performance*. New York: Palgrave MacMillan.

Mahmood, Saba. 2005. *Politics of Piety: The Islamic Revival and the Feminist Subject*. Princeton: Princeton University Press.

Mallory, Chaone. 2013. "Locating Ecofeminism in Encounters with Food and Place." *Journal of Agricultural and Environmental Ethics* 26 (1). DOI: 10.1007/s10806-011-9373-8.

Mander, Jerry, and Edward Goldsmith, eds. 1996. *The Case Against the Global Economy and For a Turn Toward the Local*. San Francisco: Sierra Club Books.

Manning, Richard. 2005. *Against the Grain: How Agriculture Has Hijacked Civilization*. New York: North Point Press.

Măntescu, Livu. 2016. "Ecoporn, Irrationalities and Radical Environmentalism." THESys Discussion Paper No. 2016-3. Humboldt University, Berlin.

Marczak, Michal. 2012. *Fuck For Forest*. Film Documentary.

Marks, Robert. 2019. *The Origins of the Modern World: A Global and Environmental Narrative from the Fifteenth to the Twenty-First Century* (4th Edition). Lanham: Rowman & Littlefield.

Marocco, Joe. 2008. "Climate Change and the Limits of Knowledge." In *The Virtues of Ignorance: Complexity, Sustainability, and the Limits of Knowledge*, edited by Bill Vitek and Wes Jackson, 307–322. Lexington: University Press of Kentucky.

Marovich, Beatrice. 2017. "The Trouble with Commonality: Theology, Evolutionary Theory, and Creaturely Kinship." In *Entangled Worlds: Religion, Science, and New Materialisms*, edited by Catherine Keller and Mary Jane Rubenstein, 317–330. New York: Fordham University Press.

Martin, Craig. 2012. *A Critical Introduction to the Study of Religion*. Bristol: Equinox Publishing.

Martin, Joel. 2001. *The Land Looks After Us: A History of Native American Religion*. New York: Oxford University Press.

Mason, Jim. 2006. "The Animal Question: Uncovering the Roots of Our Domination of Nature and Each Other." In *Igniting a Revolution: Voices in Defense of the Earth*, edited by Steven Best and Anthony Nocella, II, 178–185. Oakland: AK Press.

McCoy, Jack. 2014. *A Deeper Shade of Blue*. Indie Rights. [documentary film].

McCutcheon, Russell. 1997. *Manufacturing Religion: The Discourse on Sui Generis Religion and the Politics of Nostalgia*. New York: Oxford University Press.

McDuff, Mallory. 2010. *Natural Saints: How People of Faith Are Working to Save God's Earth*. New York: Oxford University Press.

McFarland Taylor, Sarah. 2019. *Ecopiety: Green Media and the Dilemma of Environmental Virtue*. New York: NYU Press.

McGinnis, Michael, ed. 1998. *Bioregionalism*. New York: Routledge.

McHugh, James. 2012. *Sandalwood and Carrion: Smell in Indian Religion and Culture*. New York: Oxford University Press.

McIntosh, Alastair. 2008. *Hell and Highwater: Climate Change, Hope and the Human Condition*. Edinburgh: Birlinn.

McIntosh, Alastair. 2020. *Riders on the Storm: The Climate Crisis and the Survival of Being*. Edinburgh: Birlinn.

McKibben, Bill. 2010. *Eaarth: Making a Life on a Tough New Planet*. New York: St. Martin's Griffin.

Merchant, Carolyn. 1980. *The Death of Nature: Women, Ecology and the Scientific Revolution*. San Francisco: HarperSanFrancisco.

Mingle, Jonathan. 2015. *Fire and Ice: Soot, Solidarity, and Survival on the Roof of the World*. New York: St. Martin's Press.

Mitchell, Timothy. 2011. *Carbon Democracy: Political Power in the Age of Oil*. London: Verso.

Mohorčich, Joseph. 2020. "Energy Intensity and Human Mobility After the Anthropocene." *Sustainability* 12: 2376.

Montano, Linda, Annie Sprinkle, and Veronica Vera. 1989. "Summer Saint Camp 1987: With Annie Sprinkle and Veronica Vera." *TDR (1988-)* 33 (1): 94–103.

Moran, Emilio. 1996. "Nurturing the Forest: Strategies of Native Amazonians." In *Redefining Nature: Ecology, Culture and Domestication*, edited by Roy Ellen and Katsuyoshi Fukui, 531–556. Washington, D.C.: Berg.

Morrison, Kenneth. 2000. "The Cosmos as Intersubjective: Native American Other-Than-Human Persons." In *Indigenous Religions: A Companion*, edited by Graham Harvey, 23–36. New York: Bloomsbury.

Mortimer-Sandilands, Catriona, and Bruce Erickson. 2010. *Queer Ecologies: Sex, Nature, Politics, Desire*. Bloomington: Indiana University Press.

Morton, Tim. 2009. *Ecology Without Nature: Rethinking Environmental Aesthetics*. Cambridge: Harvard University Press.

Morton, Timothy. 2010a. "Guest Column: Queer Ecology." *Proceedings of the Modern Language Association of America* 125 (2): 273–282.

Morton, Timothy. 2010b. *The Ecological Thought*. Cambridge: Harvard University Press.

Morton, Timothy. 2013. *Hyperobjects: Philosophy and Ecology After the End of the World*. Minneapolis: University of Minnesota Press.

Morton, Timothy. 2017. *Humankind: Solidarity with Nonhuman People*. New York: Verso.

Moser, Patrick. 2016. "The Endurance of Surfing in 19t-century Hawai'i." *Journal of the Polynesian Society* 125 (4): 411–432.

Mounk, Yascha. 2018. "What An Audacious Hoax Reveals About Academia." *The Atlantic*. https://www.theatlantic.com/ideas/archive/2018/10/new-sokal-hoax/572212/?fbclid=IwAR2ry7c96jPh4MTgSKC1xeHiM3X21kvZ2sHiMbb89GGluPJorGXVoqRQYW.

Munoz, Jose Esteban. 2015. "Theorizing Queer Inhumanisms." *GLQ: A Journal of Lesbian and Gay Studies* 21 (2–3): 209–210.

Nardini, Krizia. 2014. "Becoming Otherwise: Embodied Thinking and the 'Transformative Matter' of (New) Feminist Materialist Theorizing." *Artnodes: E-Journal on Art, Science and Technology* 14: 18–24.

National Research Council. 2010. *Understanding Climate's Influence on Human Evolution*. Washington, D.C.: National Academies Press.

Neimanis, Astrida, and Rachel Loewen Walker. 2014. "*Weathering*: Climate Change and the 'Thick Time' of Transcorporeality." *Hypatia* 29 (3): 558–575.

Nelson, Lance. 2006. "Cows, Elephants, Dogs, and Other Lesser Embodiments of *Ātman*: Reflections on Hindu Attitudes Toward Nonhuman Animals." In *A Communion of Subjects: Animals in Religion, Science & Ethics*, edited by Paul Waldau and Kimberley Patton, 179–193. New York: Columbia University Press.

Norberg-Hodge, Helena. 2009. *Ancient Futures: Lessons from Ladakh for a Globalizing World*. San Francisco: Sierra Club Books.

Northcott, Michael. 2015. *Place, Ecology and the Sacred: The Moral Geographies of Sustainable Communities*. New York: Bloomsbury Academic.

Nüsser, Marcus, and Ravi Baghel. 2015. "The Emergence of the Cryoscape: Contested Narratives of Himalayan Glacier Dynamics and Climate Change." In *Environmental and Climate Change in South and Southeast Asia: How Are Local Cultures Coping?*, edited by Barbara Schuler, 138–156. Leiden: Brill.

Orr, David. 2016. *Dangerous Years: Climate Change, the Long Emergency, and the Way Forward*. New Haven: Yale University Press.

Orsi, Robert, ed. 1999. *Gods of the City: Religion and the American Urban Landscape*. Bloomington: Indiana University Press.

Paavolainen, Teemu. 2018. *Theatricality and Performativity: Writings on Texture from Plato's Cave to Urban Activism*. Switzerland: Palgrave MacMillan.

Patterson, Bobbi, and Robert Puckett. 2015. "Resilience, Lived Scholarship, and Sustainable Life." *Journal of the American Academy of Religion* 83 (2): 407–421.

Petrocultures Research Group. 2016. *After Oil*. Edmonton: Petrocultures Research Group.

Plumwood, Val. 1993. *Feminism and the Mastery of Nature*. New York: Routledge.

Posey, Darrell Addison, ed. 1999. *Cultural and Spiritual Values of Biodiversity*. London: Intermediate Technology Publications.

Prager, Brad. 2017 "German Film Ventures into the Amazon: Werner Herzog's *Fitzcarraldo* as Prelude to Michal Marczak's Eco-documentary." In *German Ecocriticism in the Anthropocene*, edited by Caroline Schaumann and Heather Sullivan, 229–245. New York: Palgrave Macmillan.

Purdy, Jedediah. 2015. *After Nature: A Politics for the Anthropocene*. Cambridge: Harvard University Press.

Purdy, Jedediah. 2019. *This Land is Our Land: The Struggle for a New Commonwealth*. Princeton: Princeton University Press.

Rappaport, Roy. 1979. *Ecology, Meaning, and Religion*. Berkeley: North Atlantic Books.

Rasmussen, Larry. 2013. *Earth-Honoring Faith: Religious Ethics in a New Key*. New York: Oxford University Press.

Raworth, Kate. 2012. "A Safe and Just Space for Humanity: Can We Live Within the Doughnut?" Oxfam Discussion Paper (February). https://www-cdn.oxfam.org /s3fs- public/file_attachments/dp-a-safe-and-just-space-for-humanity-130212-en _5.pdf.

Ray, Sarah. 2020. *A Field Guide to Climate Anxiety: How to Keep Your Cool on a Warming Planet*. Berkeley: University of California Press.

Rebenstich, Juliane. 2013. "Rousseau's Heterotopology of the Theatre." In *Performance and the Politics of Space: Theatre and Topology*, edited by Erika Fischer-Lichte and Benjamin Wihstutz, 142–165. New York: Routledge.

Renner, Michael. 2013. "Climate Change and Displacements." In *State of the World 2013: Is Sustainability Still Possible?*, edited by Lester Brown, 343–352. Washington, D.C.: Island Press.

Richards, Paul. 1996. "Agrarian Creolization: The Ethnobiology, History, Culture and Politics of West African Rice." In *Redefining Nature: Ecology, Culture and Domestication*, edited by Roy Ellen and Katsuyoshi Fukui, 291–318. Washington, D.C.

Richardson, Niall. 2014. *Body Studies: The Basics*. London: Routledge.

Roberts, J. Timmons. 2007. "Globalizing Environmental Justice." In *Environmental Justice and Environmentalism: The Social Justice Challenge to the Environmental Movement*, edited by Ronald Sandler and Phaedra Pezzullo, 285–308. Cambridge: The MIT Press.

Rocha, Juan, Garry Peterson, Örjan Bodin, and Simon Levin. 2018. "Cascading Regime Shifts Within and Across Scales." *Science* 362: 1379–1383.

Rockström, Johan, Will Steffen, Kevin Noone, Åsa Persson, F. Stuart Chapin, Eric F. Lambin, Timothy M. Lenton, M. Scheffer, C. Folke, H. J. Schellnhuber, and B. Nykvist. 2009. "A Safe Operating Space for Humanity." *Nature* 461: 472–475.

Rodrigues, Hillary. 2011. "Introduction." In *Studying Hinduism in Practice*, 1–9. New York: Routledge.

Roughgarden, Joan. 2009. *Evolution's Rainbow: Diversity, Gender, and Sexuality in Nature and People*. Berkeley: University of California Press.

Sandermana, Jonathan, Tomislav Hengle, and Gregory Fiskeaw. 2017. "Soil Carbon Debt of 12,000 Years of Human Land Use." *Proceedings of the National Academy of Sciences of the United States of America* 114 (36): 9575–9580.

Sax, William. 1991. *Mountain Goddess: Gender and Politics in a Himalayan Pilgrimage*. New York: Oxford University Press.

Schade, Leah, and Margaret Bullitt-Jonas, eds. 2019. *Rooted & Rising: Voices of Courage in a Time of Climate Crisis*. Lanham: The Rowman & Littlefield Publishing Group.

Schaefer, Donovan. 2015. *Religious Affects: Animality, Evolution, and Power*. Durham: Duke University Press.

Scheiner, Samuel, and Michael Willig. 2011. "A General Theory of Ecology." In *The Theory of Ecology*, edited by Samuel Sheiner and Michael Willig, 3–20. Chicago: The University of Chicago Press.

Schendler, Auden. 2009. *Getting Green Done: Hard Truths from the Front Lines of the Sustainability Revolution*. New York: Public Affairs.

Selengut, Charles. 2017. *Sacred Fury: Understanding Religious Violence* (3rd Edition). Lanham: Rowman & Littlefield.

Seremetakis, C. Nadia. 1994a. "Prologue." In *The Senses Still: Perception and Memory as Material Culture in Modernity*, edited by C. Nadia Seremetakis, vii–xi. San Francisco: Westview Press.

Seremetakis, C. Nadia. 1994b. *The Senses Still: Perception and Memory as Material Culture in Modernity*. San Francisco: Westview Press.

Sharma, Arjun. 2019. "Giving Water Its Place: Artificial Glaciers and the Politics of Place in a High-Altitude Himalayan Village." *Water Alternatives* 12 (3): 993–1016.

Shiva, Vandana. 2009. "Women and the Gendered Politics of Food." *Philosophical Topics* 37 (2): 17–32.

Shiva, Vandana. 2016. *Staying Alive: Women, Ecology, and Development*. Berkeley: North Atlantic Books.

Shresta, Arun, and Raju Aryal. 2011. "Climate Change in Nepal and Its Impact on Himalayan Glaciers." *Regional Environmental Change* 11: S65–S77.

Sideris, Lisa. 2017 *Consecrating Science: Wonder, Knowledge, and the Natural World*. Berkeley: University of California Press.

Sideris, Lisa. 2020. "Grave Reminders: Grief and Vulnerability in the Anthropocene." *Religions* 11: 293.

Simpson, Paul. 2015. "What Remains of the Intersubjective?: On the Presencing of Self and Other." *Emotion, Space and Society* 14: 65–73.

Slater, Thomas, Isobel R. Lawrence, Inès N. Otosaka, Andrew Shepherd, Noel Gourmelen, Livia Jakob, Paul Tepes, Lin Gilbert, and Peter Nienow. 2021. "Review Article: Earth's Ice Imbalance." *The Cryosphere* 15: 233–246.

Slingerland, Edward. 2008. *What Science Offers the Humanities: Integrating Body and Culture*. New York: Cambridge University Press.

Snarey, John. 1996. "The Natural Environment's Impact Upon Religious Ethics: A Cross-Cultural Study." *Journal for the Scientific Study of Religion* 35 (2): 85–96.

Solnit, Rebecca. 2005. *Hope in the Dark: The Untold History of People Power*. New York: Canongate.

Soper, Kate. 1995. "Feminism and Ecology: Realism and Rhetoric in the Discourses of Nature." *Science, Technology and Human Values* 20: 311–331.

Sowers, Jeannie, Avner Vengosh, and Erika Weinthal. 2011. "Climate Change, Water Resources, and the Politics of Adaptation in the Middle East and North Africa." *Climatic Change* 104: 599–627.

Spencer, Daniel. 2018. "Introduction: Religion, Nature, and Queer Theory." In *Unsettling Science and Religion: Contributions and Questions from Queer Studies*, edited by Whitney Bauman and Lisa Stenmark, 15–21. New York: Lexington Books.

Sprinkle, Annie, and Elizabeth Stephens. 2012. "On Becoming Appalachian Moonshine." *Performance Research* 17 (4): 61–66.

Sprinkle, Annie, and Elizabeth Stephens. 2017/2019. *Water Makes Us Wet: An Ecosexual Adventure*. Juno Productions.

Srinivasan, Bina. 2012. "The Taming of a River: Gender, Displacement and Resistance in Anti-Dam Movements." *Peace Prints: South Asian Journal of Peacebuilding* 4 (1).

Stevenson, Jill. 2013. *Sensational Devotion: Evangelical Performance in Twenty-First Century America*. Ann Arbor: University of Michigan Press.

Stibbe, Arran. 2012. *Animals Erased: Discourse, Ecology, and Reconnection with the Natural World*. Middletown: Wesleyan University Press.

Stoekl, Allan. 2014. "Foreword." In *Oil Culture*, edited by Ross Barrett and Daniel Worden, xi–xiv. Minneapolis: University of Minnesota Press.

Stoller, Paul. 1997. *Sensuous Scholarship*. Philadelphia: University of Pennsylvania Press.

Strenski, Ivan. 2014. "Can Religion Professors Save the Planet?" *Religion Dispatches*. http://religiondispatches.org/can-religion-professors-save-the-planet/.

Swearer, Donald, and Susan McGarry, eds. 2011. *Ecologies of Human Flourishing*. Cambridge: Center for the Study of World Religions.

Szeman, Imre. 2007. "System Failure: Oil, Futurity, and the Anticipation of Disaster." *South Atlantic Quarterly* 106 (4): 805–823.

Szeman, Imre. 2013. "How to Know about Oil: Energy Epistemologies and Political Futures." *Journal of Canadian Studies* 47 (3): 145–168.

Szeman, Imre, and Dominic Boyer. 2017. *Energy Humanities: An Anthology*. Baltimore: Johns Hopkins University Press.

Szeman, Imre, and Maria Whiteman. 2012. "Oil Imag(e)inaries: Critical Realism and the Oil Sands." *Imaginations: Journal of Cross-Cultural Image Studies* 3 (2): 46–67.

TallBear, Kim. 2011. "Why Interspecies Thinking Needs Indigenous Standpoints." *Theorizing the Contemporary Cultural Anthropology Website*.

TallBear, Kim. 2015. "An Indigenous Reflection on Working Beyond the Human/Not Human." *GLQ: A Journal of Lesbian and Gay Studies* 21 (2–3): 230–235.

Taves, Ann. 2011. *Religious Experience Reconsidered: A Building-Block Approach to the Study of Religion and Other Special Things*. Princeton: Princeton University Press.

Taylor, Affrica, Veronica Pacini-Ketchabaw, Sandrina de Finney, and Mindy Blaise. 2015. "Inheriting the Ecological Legacies of Settler Colonialism." *Environmental Humanities* 7: 129–132.

Taylor, Bron. 1995. *Ecological Resistance Movements: The Global Emergence of Radical and Popular Environmentalism*. Albany: SUNY Press.

Taylor, Bron. 2005. *The Encyclopedia of Religion and Nature*. New York: Thoemmes Continuum.

Taylor, Bron. 2007. "Surfing into Spirituality and a New, Aquatic Nature Religion." *Journal of the American Academy of Religion* 75 (4): 923–951.

Taylor, Bron. 2010. *Dark Green Religion: Nature Spirituality and the Planetary Future*. Berkeley: University of California Press.

Taylor, Bron. 2016a. "The Greening of Religion Hypothesis (Part One): From Lynn White, Jr. and Claims that Religions Can Promote Environmentally Destructive Attitudes and Behaviors to Assertions they are Becoming Environmentally Friendly." *Journal for the Study of Religion, Nature and Culture* 10 (3): 268–305.

Taylor, Bron. 2016b. "Natural Religion: Nature, Science, and Religion." In *Religion: Sources, Perspectives, and Methodologies*, edited by Jeffrey Kripal. Farmington Hills: Macmillan Reference.

Taylor, Bron, Jen Wright, and Todd LeVasseur. 2020. "Dark Green Humility: Religious, Psychological, and Affective Attributes of Proenvironmental Behaviors." *Journal of Environmental Studies and Sciences* 10: 41–56.

Thayer, Robert. 2003. *Life Place: Bioregional Thought and Practice*. Berkeley: University of California Press.

Theisen, Ole Magnus. 2008. "Blood and Soil? Resource Scarcity and Internal Armed Conflict Revisited." *Journal of Peace Research* 45 (6): 801–818.

Thiele, Leslie Paul. 1999. "Evolutionary and Ecological Ethics." *Political Theory* 27 (1): 6–38.

Thiele, Leslie Paul. 2011. *Indra's Net and the Midas Touch: Living Sustainably in a Connected World*. Cambridge: The MIT Press.

Toadvine, Ted. 2009. *Merleau-Ponty's Philosophy of Nature*. Evanston, IL: Northwestern University Press.

Todd, Zoe. 2016. "An Indigenous Feminist's Take on the Ontological Turn: 'Ontology' Is Just Another Word for Colonialism." *Journal of Historical Sociology* 29 (1): 4–22.

Tola, Miriam. 2019. "Planetary Lovers: On Annie Sprinkle and Beth Stephen's *Water Makes us Wet*." In *Other Globes: Past and Peripheral Imaginations of Globalization*, edited by Simon Ferdinand, Irene Villaescusa-Illan, and Esther Peer, 231–248. Palgrave Studies in Globalization, Culture and Society. New York: Palgrave Macmillan.

Tomalin, Emma. 2002. "The Limitations of Religious Environmentalism for India." *Worldviews: Global Religions, Culture, and Ecology* 6: 12–30.

Tomalin, Emma. 2013. *Biodivinity and Biodiversity: The Limits to Religious Environmentalism*. Burlington: Ashgate.

Tuan, Yi-Fu, and Ti-Fu Tuan. 1970. "Our Treatment of the Environment in Ideal and Actuality: A Geographer Observes Man's Effect on Nature in China and in the Pagan and Christian West." *Sigma Xi, The Scientific Research Honor Society* 58 (3): 244–249.

Turner, Colin. 2011. *Islam: The Basics* (2nd Edition). New York: Routledge.

Turner, Nancy, and Helen Clifton. 2009. "'It's So Different Today': Climate Change and Indigenous Lifeways in British Columbia, Canada." *Global Environmental Change* 19: 180–190.

Turner, Victor. 1967. *The Forest of Symbols: Aspects of Ndembu Ritual*. Ithaca: Cornell University Press.

Tweed, Thomas. 1997. "Introduction: Narrating U.S. Religious History." In. *Retelling U.S. Religious History*, edited by Thomas Tweed, 1–23. Berkeley: University of California Press.

Tweed, Thomas. 2008. *Crossing and Dwelling: A Theory of Religion*. Cambridge: Harvard University Press.

van Dooren, Thom. 2014. *Flight Ways: Life and Loss at the Edge of Extinction*. New York: Columbia University Press.

Van Dooren, Thom. and Deborah Bird Rose. 2017. "Lively Ethography: Storying Animist Worlds." In *Environmental Humanities: Voices from the Anthropocene*, edited by Serpil Oppermann and Serenella Iovino, 255–271. London: Rowman & Littlefield International.

Vásquez, Manuel. 2011. *More Than Belief: A Materialist Theory of Religion*. New York: Oxford University Press.

Viveiros De Castro, Eduardo. 2015. *The Relative Native: Essays on Indigenous Conceptual Worlds*. Chicago: Hau Books.

Vries, P. H. H. 2001. "Are Coal and Colonies Really Crucial? Kenneth Pomeranz and the Great Divergence." *Journal of World History* 12 (2): 407–446.

Waitt, Gordon. 2008. "'Killing Waves': Surfing, Space and Gender." *Social & Cultural Geography* 9 (1): 75–94.

Walker, Isaiah Helekunihi. 2015. *Waves of Resistance: Surfing and History in Twentieth Century Hawai'i*. Honolulu: University of Hawai'i Press.

Wallace, MEgan. 2019. "Eco-Porn: The Movement That Says Sex Can Save The Planet." *Dazed & Confused: A Future World: Life & Culture*, May 14, 2019. https ://www.dazeddigital.com/life-culture/article/44391/1/ecosexuals-eco-porn-sex-env ironmentalism-elizabeth-stephens-annie-sprinkle.

Walsh, Lynda. 2013. *Scientists as Prophets: A Rhetorical Genealogy*. New York: Oxford University Press.

Warren, Karen. 1990. "The Power and Promise of Ecological Feminism." *Environmental Ethics* 12 (2): 125–146.

Watson, Annette, and Orville Huntington. 2008. "They're *Here*—I Can *Feel* Them: The Epistemic Spaces of Indigenous and Western Knowledges." *Social & Cultural Geography* 9 (3): 257–281.

Weadick, Paul. 2013. "Queering the Environment: Annie Sprinkle Loving the Earth (...In *That* Way)." *Hard Wire: The Undergraduate Journal of Sexual Diversity Studies* 1: 89–101.

Weaver, Jace. 2015. "Misfit Messengers: Indigenous Religious Traditions and Climate Change." *Journal for the American Academy of Religion* 83 (2): 320–335.

Weber, Andreas. 2017. *Matter and Desire: An Erotic Ecology*. White River Junction: Chelsea Green Publishing.

Wexler, J. 2016. *When God Isn't Green: A World-wide Journey to Places Where Religious Practice and Environmentalism Collide*. Boston: Beacon Press.

Whitehouse, Harvey. 2007. "Towards an Integration of Ethnography, History, and the Cognitive Science of Religion." In *Religion, Anthropology, and Cognitive Science*, edited by Harvey Whitehouse and James Laidlaw, 247–280. Durham: Carolina Academic Press.

Williams, Joseph, Mohammed Shobrak, Thomas M. Wilms, Ibrahim A. Arif, and Haseeb A. Khan. 2012. "Climate Change and Animals in Saudi Arabia." *Saudi Journal of Biological Sciences* 19: 121–130.

Williams, Linda. 1993. "A Provoking Agent: The Pornography and Performance Art of Annie Sprinkle." *Social Text* 37: 117–133.

Wilson, E. O. 2006. *The Creation: An Appeal to Save Life on Earth*. New York: W.W. Norton & Company.

Wilson, Sheena, Imre Szeman, and Adam Carlson. 2017. "On Petrocultures: Or, Why We Need to Understand Oil to Understand Everything Else." In *Petrocultures: Oil, Politics, Culture*, edited by Sheena Wilson, 3–19. McGill: Queen's University Press.

Winkelman, Michael, and John Baker. 2010. *Supernatural as Natural: A Biocultural Approach to Religion*. Upper Saddle River: Pearson Prentice Hall.

Wohl, Ellen. 2011. *Environmental Change on Ten of the World's Great Rivers*. Chicago: The University of Chicago Press.

Worster, Donald. 2015. "A Long Cold View of History: How Ice, Worms and Dirt Made Us What We Are Today." *The American Scholar*. Spring. https://theamer icanscholar.org/a-long-cold-view-of-history/#.WrKa-uch3IUv.

Xu, Jianchu, R. Edward Grumbine, Arun Shrestha, Mats Eriksson, Xuefei Yang, Y. U. N. Wang, and Andreas Wilkes. 2009. "The Melting Himalayas: Cascading Effects of Climate Change on Water, Biodiversity, and Livelihoods." *Conservation Biology* 23 (3): 520–530.

Yarrow, Ralph, ed. 2007. *Sacred Theatre*. Bristol: Intellect.

Yonnetti, Eben. 2020, November 2. "How Buddhist Monks in Ladakh Are Fighting Climate Change." *Lion's Roar: Buddhist Wisdom for our Time*. https://www w.lionsroar.com/how-buddhist-monks-in-ladakh-are-fighting-climate-change/. Accessed May 26, 2020.

Yusuff, Kathryn. 2018. *A Billion Black Anthropocenes or None*. Minneapolis: University of Minnesota Press.

Zylinska, Joanna. 2018. *The End of Man: A Feminist Counterapocalypse*. Minneapolis: University of Minnesota Press.

Index

academy/the academy/higher education (during climate change), xv. xvii, xx–xxi, xxviii–xxix, xxx, 8, 28–29, 64, 65, 69–71, 114, 136, 143, 144, 150–53; education for flourishing during climate change, 150–55, 157

agency, xxv, xxix, xxxi, 27, 53, 64, 65, 80, 86, 144, 157; sympoietic/sympoiesis, xxx, xxxi, 18, 29, 37, 44, 66, 80, 84, 115, 124, 152

Anthropocene, xix, xxvii, xxix, xxx, 62, 65, 69, 70, 72, 147, 153–55; alternative names to, 41. *See also* Carboncene; Symbiocene; criticisms of the term, xix–xx, 140; post-Anthropocene, 63, 64; violence from/of, xx, 61, 64, 71–72, 97, 115

biocultural, xiv, 11, 19, 26, 38. 48, 54, 70, 77, 83, 98, 99, 101–2, 122, 141, 144. 152; definition of/naturalistic understanding of, 24–25, 80

bio-ecologies/ecologies of place, xiv, xxi–xxii, 4, 11–12, 19, 26, 49, 54, 59, 60, 63–65, 69, 70, 77, 79, 84, 86, 97, 99, 116, 119, 144, 145, 150, 152, 157; academic cultures of and praxis of resilience, 152–55; cultures of (communication, language,

symbolism), 39, 52, 53; dwelling within, 37–38, 152. *See also* Ingold, Tim; nature, 43, 45–46

bodies, 25–26, 51, 152. *See also* chapters 5, 6 and 7; Hindu views of; embodiment, xiv, 80, 111, 124; erotic, 124; religious, xiii, 26, 60; theories of, xvi, 4

Butler, Judith, 5–6, 124

Carboncene, 71, 77, 83, 115

climate change(s)/climate warming/global warming, xiv, xv–xvi, xviii–xix, xxii, 8–13, 15, 39, 44, 52, 53, 69–71, 74, 80, 86, 99, 102, 111, 119, 124, 143, 149–50; agriculture, role in, 9–10, 15, 61, 62; in the Arabian Peninsula, 88–89; biogeochemical, xxiv, 3–4, 9, 11, 15, 30–31, 44, 70, 72, 77, 79, 87, 99, 150, 152; in the Himalayas, 82–83, 99–100; soils, 11. *See also* performance; petroculture

COVID-19; economic and environmental impact of, 12–13; religious responses to, 13, 75

dualisms, xv, xviii, xxi, 21, 145; abstraction from place, 23–24; critiques of/critiques of